T0134901

Cyber-Physical System Design from an Architecture Analysis Viewpoint

Shin Nakajima · Jean-Pierre Talpin
Masumi Toyoshima · Huafeng Yu
Editors

Cyber-Physical System Design from an Architecture Analysis Viewpoint

Communications of NII Shonan Meetings

 Springer

Editors
Shin Nakajima
National Institute of Informatics
Tokyo
Japan

Jean-Pierre Talpin
Inria
Rennes
France

Masumi Toyoshima
DENSO Corporation
Kariya, Aichi
Japan

Huafeng Yu
Boeing Research & Technology
Huntsville, AL
USA

ISBN 978-981-13-5136-5   ISBN 978-981-10-4436-6   (eBook)
DOI 10.1007/978-981-10-4436-6

Printed on acid-free paper

This Springer imprint is published by Springer Nature
The registered company is Springer Nature Singapore Pte Ltd.
The registered company address is: 152 Beach Road, #21-01/04 Gateway East, Singapore 189721, Singapore

Preface

The term cyber-physical system (CPS) was introduced by Helen Gill at the NSF referring to the integration of computation and physical processes. In CPS, embedded computers and networks monitor and control the physical processes, usually with feedback loops where physical processes affect computations and vice versa. The principal challenges in system design lie in this perpetual interaction of software, hardware, and physics.

CPS safety is often critical for society in many applications such as transportations, whether automotive, trains or airplanes, power distribution, medical equipment, or tele-medicine. Whether or not life is threatened, failures may have huge economic impact. Developing reliable CPS has become a critical issue for the industry and society. Safety and security requirements must be ensured by means of strong validation tools. Satisfying such requirements including quality of service implies to have the required properties of the system proved formally before it is deployed.

In the past 15 years, technologies have moved towards Model Driven Engineering (MDE). With the MDE methodology, requirements are gathered with use cases, and then a model of the system is built that satisfies the requirements. Among several modeling formalisms appeared in these years, the most successful are *executable* models, models that can be exercised, tested, and validated. This approach can be used for both software and hardware.

A common feature of CPS is the predominance of concurrency and parallelism in models. Research on concurrent and parallel systems has been split into two different families. The first is based on synchronous models, primarily targeting design of hardware circuits and/or embedded and reactive systems. Esterel, Lustre, Signal, and SCADE are examples of existing technologies of this nature. Additionally, in many places, these have been connected with models of environments that are required for CPS modeling. The second family addresses loosely coupled systems, where communication between distributed entities is asynchronous by nature. Large systems are, indeed, mixing both families of concurrency. They are structured hierarchically and they comprise multiple synchronous devices connected by networks that communicate asynchronously.

In an architectural model, a CPS is represented by a distributed system. The model consists of components with well-defined interfaces and connections between ports of the component interfaces. Furthermore, it specifies component properties that can be used in analytical reasoning about the system. Models are hierarchically organized: each component can contain another sub-system with its own set of components and connections between them. An architecture description language for embedded systems, for which timing and resource availability are primary characteristics of requirements, must additionally describe resources of the system platform, such as processors, memories, and communication links. Several architectural modeling languages for embedded systems have emerged in recent years, including the SAE AADL, SysML, and UML MARTE.

An architectural specification serves several important purposes. First, it breaks down a system model into components of manageable size to establish clear interfaces between these components. In this way, complexity becomes manageable by hiding details that are not relevant at a given level of abstraction. Clear, formally defined, component interfaces allow us to avoid integration problems at the implementation phase. Connections between components, which specify how components affect each other, help propagate the effects of a change in one component to the linked components.

More importantly, an architectural model serves as a repository to share knowledge about the system being designed. This knowledge can be represented as requirements, design artifacts, and component implementations, all of which are held together by a structural backbone. Such a repository enables automatic generation of analytical models for different aspects of the system, such as timing, reliability, security, performance, or energy consumption.

In most cases, however, quantitative reasoning in architecture modeling and CPS is predominantly parameterized by the dimension of time. An architecture or CPS model refers to software, hardware, and physics. In each of these viewpoints, time takes a different form: continuous or discrete, event-based or time-triggered. It is, therefore, of prime importance to mitigate heterogeneous notions of time to support quantitative reasoning in system design, either using a tentatively unified model for it, or by formalizing abstraction/refinement relations from one to another in order to mitigate heterogeneity.

Despite recent research activities in the aim of formally defined or semantically interpreted architectural models, we observe a significant gap between the state of the art and practical needs to handle evolving complex models. In practice, most approaches cover a limited subset of the language and target a small number of modeling patterns. A more general approach would most likely require a semantic interpretation, an abstraction and a refinement, of the source architecture model by the target analytic tool, instead of hard-coding semantics and patterns into the model generator.

Duing March 21–24, 2016, we organized an NII Shonan Meeting on *Architecture-Centric Modeling, Analysis, and Verification of Cyber-Physical Systems*. The meeting invited 22 world-leading researchers from Asia, Europe, and North America and provided a unique opportunity for the participants to exchange ideas and develop a common understanding of the subject matter. Further information about this meeting can be found at the Web site (http://shonan.nii.ac.jp/seminar/073/) and in the meeting report, No. 2016-5 (http://shonan.nii.ac.jp/shonan/report/).

The meeting brought together contributions by research groups interested in defining precise semantics for architecture description languages, and of using this mathematical foundation to leverage tooled methodologies. Such new methodologies generate analytical models, for the purpose of simulation and formal verification, components integration, or requirements validation. Furthermore, they generate code, for the purpose of interfacing components properly, for the purpose of orchestrating execution of components with heterogeneous policies, and for the purpose of real-time scheduling execution of application thread components.

The book consists of chapters addressing these issues, written by the participants of the meeting, which offers snapshots on the state of the art in each of every viewpoint of the problem at hand, and means to put them through.

Tokyo, Japan   Shin Nakajima
Rennes, France   Jean-Pierre Talpin
Kariya, Japan   Masumi Toyoshima
Huntsville, USA   Huafeng Yu
January 2017

Acknowledgements

We, the editors, would like to thank the committee of the Shonan Meeting at NII (National Institute of Informatics, Tokyo) for supporting the meeting that led to the idea for this book. We thank the meeting participants for their active and interesting discussions during the meeting and their contributions to this book. We are also grateful to the book contributors for the help in reviewing the chapters. Last but not least, we thank the staff of Springer who gave us the opportunity and support for publication of this book.

Contents

Contributors

Toshiaki Aoki Japan Advanced Institute of Science and Technology, Nomi, Ishikawa, Japan

Loïc Besnard IRISA, Rennes, France

Etienne Borde Telecom ParisTech, Paris, France

Marco Bozzano Fondazione Bruno Kessler, Povo, Trento, Italy

Harold Bruintjes Software Modeling and Verification Group, RWTH Aachen University, Aachen, Germany

Alessandro Cimatti Fondazione Bruno Kessler, Povo, Trento, Italy

Julien Delange Carnegie Mellon Software Engineering Institute, Pittsburgh, USA

Thierry Gautier INRIA, Rennes, France

Andreas Gerstlauer The University of Texas at Austin, Austin, TX, USA

Paul Le Guernic INRIA, Rennes, France

Clément Guy INRIA, Rennes, France

Fernando Herrera TEISA, GESE, Universidad de Cantabria, Santander, Spain

Jérôme Hugues Institut Supérieur de l'Aéronautique et de l'Espace, Universit de Toulouse, Toulouse, France

Vania Joloboff INRIA Bretagne Atlantique, Campus de Beaulieu, Rennes Cedex, France

Joost-Pieter Katoen Software Modeling and Verification Group, RWTH Aachen University, Aachen, Germany

Tomoji Kishi Faculty of Science and Engineering, Waseda University, Shinjuku, Tokyo, Japan

Brian Larson FDA Scholar, Kansas State University, Manhattan, KS, USA

Frédéric Mallet Université Côte d'Azur, CNRS, Inria, I3S, Sophia Antipolis, France

Thomas Noll Software Modeling and Verification Group, RWTH Aachen University, Aachen, Germany

Makoto Satoh Renesas System Design Co., Ltd., Yokohama, Kanagawa, Japan

Jean-Pierre Talpin INRIA, Rennes, France

Mitsuhiro Tani DENSO CORPORATION, Kariya, Aichi, Japan

Stefano Tonetta Fondazione Bruno Kessler, Povo, Trento, Italy

Eugenio Villar TEISA, GESE, Universidad de Cantabria, Santander, Spain

Kenro Yatake Japan Advanced Institute of Science and Technology, Nomi, Ishikawa, Japan

Chapter 1
Virtual Prototyping of Embedded Systems: Speed and Accuracy Tradeoffs

Vania Joloboff and Andreas Gerstlauer

Abstract Virtual prototyping has emerged as an important technique for the development of devices combining hardware and software, in particular for the development of embedded computer components integrated into larger cyber-physical systems with stringent performance, safety or security requirements. Because virtual prototypes allow for observing and testing the system without requiring a real hardware at hand, they make it possible to test application software and use formal methods tools to validate the system properties at early design stages. A virtual prototype is an abstraction that emulates, with more or less accuracy, the real system under design. That emulation is usually significantly slower than the real application. In this survey, we overview different virtual prototyping techniques that can be used, and the compromises that they may offer to trade-off some aspects of reality in exchange for other higher priority objectives of the project.

Keywords Virtual prototyping · System simulation · Performance estimation · Instruction set simulation · Dynamic binary translation · Host-compiled simulation · Transaction-level modeling

1.1 Introduction

Embedded and cyber-physical systems have become ubiquitous in our everyday life, including consumer electronics, transportation, control, industry automation, energy grids, medical equipment and defence systems. Their timely development has become a key to economic success in many business areas. An inherent property of embedded systems is that they combine hardware and software into a coherent apparatus that serves some dedicated function, e.g. a camera taking pictures, or a train

V. Joloboff
INRIA Bretagne Atlantique, Campus de Beaulieu, 35042 Rennes Cedex, France
e-mail: vania.joloboff@inria.fr

A. Gerstlauer (✉)
The University of Texas at Austin, Austin, TX 78712, USA
e-mail: gerstl@ece.utexas.edu

© Springer Nature Singapore Pte Ltd. 2017
S. Nakajima et al. (eds.), *Cyber-Physical System Design from an Architecture Analysis Viewpoint*, DOI 10.1007/978-981-10-4436-6_1

braking system. Also, more and more devices have turned from being standalone to networked, communicating systems, which necessitate the adjunct of communication hardware and software to the embedded platforms.

As a corollary of these evolutions, at the beginning of a project (to design a new system) the separation may not be clearly defined yet between which system functions will be performed by hardware or by software. This hardware/software partitioning may be decided based on large number of factors, including performance, flexibility and costs, as the result of an investigation and early prototyping phase. Independently of all these considerations, there is market pressure to decrease their development time and costs, despite increasing safety and security requirements for the manufacturers, for example in transportation, industrial process control or defence. Crucially, this includes both functional as well as non-functional requirements, i.e. requirements on the behavior of a system at its outputs as a function of its inputs, as well as quality attributes such as real-time performance, power consumption or reliability.

All these issues call for development tools that can help the designers to decide the optimal hardware and software combination, to validate the software functionality, to estimate overall performance, and to validate that the resulting product verifies the required properties in timely fashion. Among available design automation approaches, simulations continue to play a vital role. A simulator provides an executable, virtual model of a candidate design that allows emulating the evolution of behavior over time and under different inputs before a system is built, i.e. for candidate designs that may not yet physically exist. Given the inherently dynamic nature of complex systems today, being able to rapidly observe and evaluate their time-varying behavior at early design stages can be crucial feedback for designers. As such, virtual prototyping as presented in this chapter is a key tool, although it is certainly not the only one available.

This chapter is aimed at providing an overview and survey of issues and solutions in virtual prototyping of embedded computer systems. The main concern in all virtual prototyping is to find the right abstractions that will allow optimally navigating fundamental tradeoffs between applicability, accuracy, development costs and simulation speed. The rest of the chapter is organized as follows: After an overview of general virtual prototyping objectives and issues in Sect. 1.2, we will discuss general abstraction concepts in Sect. 1.3 followed by various approaches that have been taken in modeling different aspects and parts of embedded computer systems in Sects. 1.4 and 1.5. Finally, the chapter concludes with an outlook on remaining issues and future directions in Sect. 1.6.

1.2   Virtual Prototyping

Engineering projects aim at constructing a device, or an ensemble of devices, that performs some function. The goal is to elaborate the object, including hardware, software, and the mechanical parts, and later to manufacture this object on a production line. In the past, engineers have been making physical prototypes of such devices by creating or assembling the hardware parts, writing the software programs running on the hardware, and testing whether the prototype would satisfy the requirements. When it does not, they would construct a second prototype, and so on, until the device is ready for manufacturing. This is a long and tedious process, slowing down time to market, and because the number of actual prototypes is limited, only a handful of engineers can work concurrently.

Computer modeling technologies in 3D, mechanics and electronics have become powerful enough that one can build a *virtual prototype*, that is, a virtual device that can be tested and validated in simulation. In this chapter, we use the term *virtual prototyping* to refer to technology that provides an emulation of an embedded computer system represented by executable hardware and software models that capture most, if not all, of the required properties of a final system. The real application software can be run over the virtual prototype of the hardware platform and produce the same results as the real device with the same sequence of outputs. The virtual prototype of the complete system under design can then be run and tested like the real one, where engineers can exercise and verify the desired device properties. However, a key challenge is that the virtual prototype includes enough detail to faithfully and accurately replicate the behavior of the modeled system. This in turn directly influences the speed at which the prototype can be run. Ideally, it should do so in about the same time frame as the real system and also report non-functional properties such as performance measurements, power consumption, thermal dissipation, etc.

The goal of virtual prototyping is to verify that required functional and non-functional properties of the system are satisfied. In a number of cases, it happens that verifying properties on a virtual platform is more advantageous than on the real hardware. This includes:

Observing and checking system properties   Sometimes one would like to observe the system internally without disturbing the device functions. In a number of cases, adding software to observe the system modifies its behavior. For example, to test that the memory usage of an embedded application is correct, one has to track the memory operations using instrumented code and this introduces changes in the application behavior, and possibly also the real target platform cannot support such instrumentation. A virtual prototype makes it possible to add non-intrusive *software probes* in the simulation model itself. This can include observation of hardware behavior, e.g. to monitor the memory consumption as each memory access can be tracked. Also, many software defects stem from hardware not being correctly initialized. Whereas it is hard to check this on the real platform, it is easy to verify initialization assertions in a virtual platform. Finally, in a virtual simulation, time can be stopped to observe the global system at a particular point in execution. This is difficult if not impossible to achieve in the physical world. Even if one system component is stopped, e.g. in a debugger, other components and the external environment will continue to run.

Accelerating time to market and lowering costs   Building physical prototypes can be extremely expensive, especially when considering prohibitive costs of taping out real chips. As such, when using real platforms, there are usually only a small number of hardware prototypes available, which severely limits the number of engineers working on the project. In contrast, one may have many engineers each running the virtual prototype on their personal workstation. Moreover, a virtual prototype is typically much cheaper to build and modify. The process of downloading the software into the target and instrumenting the tests is avoided. As many engineers as necessary can be assigned to work on the virtual prototype. A real prototype is only necessary at the very end of the process to undergo the final tests.

Complementing formal model verification   Since a virtual prototype is based on a high-level abstraction model, it is possible to use model verification tools, such as a model checker, a deadlock verifier, an invariant verifier, theorem provers, etc. to check that this model is sound. As verification is highly time consuming, it can be complemented with and aided by simulations of the same model. For example, it may be pretty difficult to reconstruct the cases where the verification tool fails. One may instead use the simulator to debug the faulty situations. Conversely, simulations can be used to generate symbolic information usable by formal methods tools. During the simulation, the system state may be observed whether in or hardware or software, and one can verify assertions, or generate symbolic trace files that can be anlayzed with formal method tools.

Checking fault tolerance and safety   Many critical systems have to be tolerant to hardware failures using redundancy or other techniques. Some faults can easily be checked in the real world, e.g. removing a cable, but some others, such as transient failures, can be very hard to create in a real experimental context. A virtual prototype can simulate more easily a defective component, whether permanently or transiently, and the fault tolerance can be checked in the virtual environment. Similarly, a virtual prototype can be easily exercised with inputs outside of normal operating conditions to check behavior under corner cases without running the risk of damaging the physical device or its environment.

In order to discuss various aspects of virtual prototyping, let us consider here the case of the iPhone virtual prototype freely distributed by the manufacturer, shown in Fig. 1.1. The iPhone Software Development Kit provides an environment in which software developers can develop and test iPhone software on a standard workstation. On the workstation display a virtual phone appears. The developers can interact with the software application under development. They can immediately test their software, without requiring any real mobile phone. No need to download the software into a real device for testing and collecting the results with some complicated mechanism. The software is run immediately, the visual output shows up on the display and resulting data may be available in local files.

This virtual prototype is a partly *functional* virtual prototype. The developers can test a lot of their application code but they cannot place a phone call. Furthermore, the iPhone simulator does not accurately simulate any hardware or operating system (OS) detail, nor does it provide any feedback about the performance of the application on the real phone. Although it is not a *full system* simulator, it has been a strong enabler of applications for the iPhone developers market.

Fig. 1.1   Popular iPhone virtual prototype

In most cases, embedded systems developers do not want to only test the functionality of their application software like iPhone developers. They want to explore hardware architectures, code new software drivers for this hardware, and build new applications for which they want to have some performance estimation. A complete virtual prototyping platform must include some operating system and hardware emulation technology since the device functions must be simulated at least to a minimum extent and with a certain amount of accuracy in order to run the software and evaluate the design alternatives. A hardware simulation engine is then a key component of the virtual prototyping platform, which makes it possible to run the application software and generate outputs that can be analyzed by other tools. This simulation engine must simulate the hardware at least enough to accurately run the software with regards to the real hardware. It is said to be *bit-accurate* when it provides identical output (to the real device) when given some input data. It is further said to be *cycle-accurate* when it provides outputs after exactly the same amount of elapsed clock cycles as the real device when given the same inputs in the same cycles.

The electronics industry has long been using electronic design automation (EDA) tools to design hardware. There are many such tools commercially available for hardware developers. They make it possible to test the design in all aspects and are fundamental to the electronics industry. EDA tools are available to simulate the hardware in every detail. This interesting property from the hardware design point of view turns out to be a drawback from the system point of view. Despite many recent advances to accelerate simulation, simulating very low level operations is much too slow to be usable for virtual prototyping. The authors of FAST [12] report that their system, although it is indeed a fast bit- and cycle-accurate simulator implemented with FPGAs, runs at 1.2 million instructions per second (MIPS), which is much faster than the systems to which they compare. Considering for example an application program that executes 60 billions instructions: it takes roughly 5 min to execute on a 200 MHz platform (assuming 1 instruction per clock cycle), but it would take over 16 h to run over a simulator at 1 MIPS.

Embedded systems designers who want to evaluate architectures and test software cannot wait for endless simulation sessions. Another approach must thus be used to provide reasonable simulation performance for quick iterative steps in modern agile development projects. In order to reduce simulation time, the virtual prototype must trade-off something in exchange for speed. Depending upon the designers' goals, one may be interested in trading some loss of accuracy, which leads to constructing simulation models that focus on some design aspects while providing abstraction of others. It also takes time to develop a virtual prototype. In many cases, developing a detailed virtual prototype from scratch would in fact take more time than assembling a real one.

In the end, some tradeoffs may have to be made between the accuracy of the virtual prototype, the time it takes to develop the virtual models, and the simulation speed. In the sequel of this chapter, we provide an overview of various abstraction techniques that can be considered for making such tradeoffs. The following terminology is used: the embedded system to be simulated is called the *target system*, and the computer system used to run the simulation is called the *host*. During a simulation session on the host some *wall clock time* is elapsing, representing some *simulated time* on the target. The faster the simulation, the larger simulated time is obtained in constant wall clock time. The simulation is *real time* when the two are equal.

1.3   Simulation Abstractions

Tradeoffs are made in general by abstracting the low-level details into higher-level functions that can be simulated with a simpler and thus faster model. However, a model at a higher level of abstraction will at the same time result in less accuracy in simulating those details that have been abstracted away. As such, there are fundamental tradeoffs between accuracy and speed when developing simulation models. A key concern in virtual prototyping is thus on deciding which aspects of a system can be abstracted away, and which parts have to be simulated in detail. These choices create a space of possible prototyping models. Ultimately, choosing the right model will depend on the particular use case and step in the design flow, where, due to inherent tradeoffs, there can never be one optimal model to satisfy all concerns. A system designer will use a range of models at successive lower levels of abstraction, gradually increasing the amount of detail as the design progresses. Fast yet relatively inaccurate models can be used at early design stages to quickly prune out unfeasible design candidates. By contrast, slow but accurate models are typically used for final validation and sign-off to physical design and manufacturing.

A key principle that permeates all of system design is the separation of largely orthogonal computation and communication aspects of a design [39]. In their overview paper, Cai and Gajski [8] have introduced a two-dimensional taxonomy of design models by applying this separation of communication from computation to the modeling space. In other words, it means that one can independently choose to abstract the low level of the computation steps or of the communication steps. Virtual prototyping models can thus provide an abstraction of the system under design in these two directions:

- *Communication abstraction* in the sense given by Cai and Gajski. Hardware components communicate together using some wiring and some protocol to transmit the data over the wiring. The communication may vary from simple one-to-one synchronous data transfer to complex asynchronous, multiplexed bus systems with priority arbitration. It may be that simulation of the communication and the particular protocol is irrelevant to the purpose of virtual prototyping. Then, part or all of the communication details can be abstracted into higher data transmission functions.
- *Computation abstraction.* A hardware component computes a high-level function by carrying out a series of steps, which in turn are executed by composing even smaller components. In a virtual prototyping environment, it is often possible to compute the high-level function directly by using the available computing resources of the simulation host machine, thus abstracting the real hardware steps.

At the same time, abstraction of both computation and communication can actually be divided in two aspects: *pure functionality* that only serves to obtain the resulting data, and *timing information* that relates to the synchronization or the delays to obtain that result. For example, simulating memory accesses with or without a cache simulation has no direct impact on the data, only on the delay to obtain it. Timing information is thereby the core non-functional performance property to be modeled and validated in a virtual prototype. Knowledge of timing provides the basis for modeling other properties such as power consumption (when combined with energy estimation), which in turn provides the input for thermal models, etc. We can thus distinguish principle abstractions along two dimensions:

- *Functional abstraction* is determined by the amount and granularity of structural detail versus purely behavioral descriptions in a model. A purely behavioral model only describes the input-output operations of a particular system or block in the form of arbitrary and fast host simulation code without any further internal detail. A structural description instead breaks down the system or block into a netlist of components, each of which can in turn be described in behavioral form or as a structural decomposition into additional sub-components. With each additional level of structural detail, additional component interactions can be observed and validated, but also additional component communication is exposed, which needs to be modeled and simulated with corresponding overhead.
- *Non-functional timing abstraction.* In a precise and accurate simulator, there are multiple parallel simulation tasks. Each task makes some progress within measured

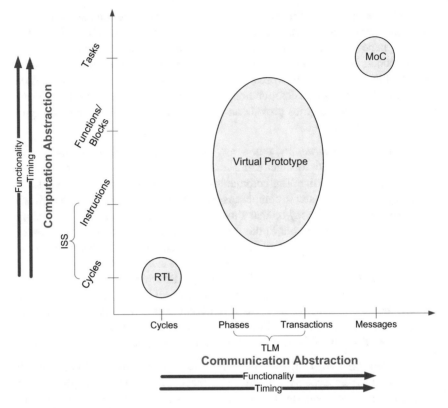

Fig. 1.2 Modeling space

clock cycles and therefore precise performance measurements can be made. But keeping track of elapsed time and synchronizing many low level tasks with the clock steps can considerably slow down the simulation. In a virtual prototype, one may not need to have that precise timing information. Instead, abstractions can be defined allowing for faster simulation, for example by ignoring cache handling and processor pipeline stalls, etc.

Functional and timing abstractions are closely coupled through the granularity of modeled computation and/or communication components. For the example of caches, in order to provide a simulation of cache delays, the memory accesses between the CPU core and the cache need to be structurally separated. Similarly, the timing impact of synchronization and concurrency effects can only be accurately captured if component interactions are faithfully modeled. Nevertheless, there are approaches that decouple functionality and timing. For example, if enough information can be extracted from a purely behavioral simulation, and if there are no complex component interactions, a fast functional simulation can be augmented with an accurate timing model without having to expose all structural detail. Such an approach can be applied to model cache delays in source-level software simulations, as will be discussed later.

In summary, the design modeling space can be viewed, as shown in Fig. 1.2, as a four-dimensional space. The coordinate origin can be viewed as the complete bit- and cycle-accurate register-transfer level (RTL) specification whereas the extreme point is an infinitely abstracted model of computation (MoC). Virtual prototyping solutions explore this modeling space to trade some characteristics in return for simulation performance. Moving into the space amounts to finding appropriate abstractions that may satisfy the prototyping requirements and yet provide reasonably fast simulation. The following two subsections review general approaches for abstracting functionality and (non-functional) timing in more detail. In the remaining sections of this chapter, we then survey the state of the art in how such abstraction approaches have been applied to computation, communication and other parts of a system model.

1.3.1   *Functional Abstraction*

As mentioned before, functional abstraction is determined by the granularity of structural detail exposed and thus observable in a model. Together with this come details of how functionality is broken down into smaller and smaller steps provided by different components and their interactions. For example, a structural model of a processor at microarchitecture granularity will describe the step-by-step process of how each instruction is processed when executing a piece of code on the processor. By contrast, a behavioral description will only emulate the equivalent functionality of the code, possibly directly at the source level without even exposing individual instructions. In the process, certain functional details and limitations of the processor microarchitecture may not be modeled and abstracted away, such as details of exception handling.

A virtual prototype is a structural description at the full system level. It models and allows observing the system architecture as a set of processing elements (PEs), such as processors, accelerators, memories or peripherals, interacting through a set of busses or other interconnect, including communication elements (CEs), such as routers, bridges and transducers. The functional abstraction of a virtual prototype is ultimately defined by the abstraction used to model its individual PE, interconnect and CE components. This in turn determines the detailed steps and events that can be observed across the system, where a larger number of steps and events negatively influence the speed at which the prototype can be simulated. We can generally distinguish four levels of abstraction:

- *Microarchitecture* simulation. The simulation of the cycle-by-cycle operation of hardware at the register-transfer level. This can either be in the form of typical hardware description languages (HDLs), such as VHDL or Verilog, or as functionally equivalent models in the form of C or similar code. HDL models can at the same time serve as the basis for further synthesis down to actual implementations.

By contrast, C models can make use of abstract data types and a hardcoded sequence of evaluations to achieve faster simulation speeds.

- *Instruction- or transaction-level* simulation. On the computation side, processor models are abstracted to only emulate the functionally equivalent behavior of each instruction. On the communication side, transaction level modeling (TLM) makes it possible to abstract complete transactions on individual phases, such as arbitration, address and data phases of read/write transactions on a bus, into function calls that emulate each transaction or phase without exposing the underlying protocol of sampling and driving various pins and wires. As will be described in subsequent sections, due to the significantly reduced detail that needs to be simulated, such models can provide orders of magnitude speedup while maintaining functional accuracy.
- *Source- or intermediate-level* simulation. Beyond instruction- or transaction-level simulation, functionality is abstracted into coarser blocks that emulate equivalent behavior without necessarily exposing individual instructions or all transactions. On the computation side, larger blocks of code or complete functions are provided in the form of source or intermediate representation (IR) code that can be compiled and executed on the simulation host to emulate their functionality on the target. On the communication side, larger message transfers consisting of multiple transactions are emulated as a single function call. Simulation speed can be further increased, but details of actual instruction or transaction streams are potentially abstracted away.
- *Purely behavioral* simulation. No structural or implementation detail is modeled, and only the overall input-output functionality of the system or subsystem is simulated in an abstract manner. In extreme cases, a MoC may only describe the inherent task-level concurrency to perform associated determinism or deadlock analyses without even modeling each task's behavior.

As already mentioned above, the granularity and level of functional abstraction is tightly related to the level of non-functional detail, chief among which is timing. The exact timing behavior of interactions, overlap and concurrency among components, such as pipeline stages in a processor microarchitecture, can in general only be accurately described if they are explicitly modeled in structural form. Nevertheless, in some cases it is possible to couple a faster behavioral description with an accurate lower-level timing model. For example, a functional simulation at the block or message level can be annotated with accurate timing information that is obtained from pre-characterization of complete code blocks or message-passing transactions on a detailed timing model.

1.3.2 Time Abstraction

Time abstraction is fundamentally coupled to the granularity at which time is simulated. A coarser granularity allows the simulator to advance time in larger steps with fewer context switches and events to be simulated, resulting in faster simulation speed. At the same time, this requires lumping multiple events into larger discrete time steps, where details about original event ordering, i.e. accuracy are lost.

A technique widely used to abstract time is so-called *temporal decoupling* or *time warping* [34]. The notion of time warping can be roughly expressed as follows: when a simulation task is activated, it must perform the operation that it has to perform now, but it might perform as well the operations it will have to carry in the future (e.g. in future clock cycles) if these operations are already known at that time (and the input data is available). The simulation task can then jump ahead of time, hence the name. An operation that takes several clock cycles may then be simulated entirely in one simulation step, whose result is available earlier than the real future clock cycle. This technique makes it possible to reduce the number of simulation tasks, reduce context switches, and possibly advance the clock by several cycles without simulating each intermediate step.

Time warping is possible only to the extent that the current operation does not need or depend on additional data or events that will be computed in the future. Two alternatives are possible. First, one can stop the simulation task until either the necessary data has become available or some relevant event has occurred. This is called *conservative simulation*. Second, one may assume in *optimistic simulation* that the necessary data will have a certain value or that no event will arrive, and continue the computation under this assumption. When the data or event indeed arrives, either the optimistic computation is determined to have been correct, or the simulation state has to be rolled-back and the operation re-computed with the correct data. If the number of such roll-backs is small, an optimistic simulation can be significantly faster than a conservative one, which can only apply time warping in cases where the model guarantees that no out-of-order events can ever occur. Both approaches are commonly applied, where tradeoffs depend on the particular context and circumstances. In addition, using the conservative approach, a simulator can be built where all the tasks carry their work as long as they have the necessary data and notify others when they produce new data. As a result, it is possible to build a purely *timeless* functional simulator with simulation tasks that only use an event structure and two primitives for waiting and notifying events. The simulation then does not use any clock. The SystemC [1, 33] modeling language makes it possible to design such timeless models using event-based synchronization, as well as timed models all the way down to cycle-accurate ones.

In the context of virtual prototyping, the granularity down to which timing needs to be measured depends on various factors. In such cases as performance evaluation or worst-case analysis, the simulation should produce accurate timing estimates. If the goal is to test and verify the hardware/software interface, then timing estimates may not be necessary. One usually distinguishes four types of time abstractions:

- *Cycle-accurate* simulation. The simulation of hardware components is implemented in software with parallel simulation tasks representing the tiny individual,
  simultaneous hardware operations. The core of the simulator can be described
  schematically as maintaining a list of tasks activated at some clock tick. On each

clock cycle, the system runs the necessary tasks, then moves to the next clock tick. This makes it possible to have a very detailed simulation, with a very precise count of clock cycles, but it is very slow.

- *Approximately timed.* Approximately timed simulators are supposed to provide accurate timings with a known error margin. Unlike cycle-accurate models, the simulation tasks may overlap several clock cycles. However, the simulation models must implement a lot of auxiliary components that are irrelevant for the functional view, slowing down the simulation. On the computation side, an approximately timed model must simulate the instruction fetch hardware and the data cache hardware and estimate the pipe line stalls in order to have an acceptable error margin. On the communication side, a model must describe the separation into different arbitration, address and data phases for each bus transaction.
- *Loosely timed* has, as the name implies, a loose definition of time. It covers a wide range of simulations where some timing information is obtained, but it is known that this information is not accurate and can possibly be far from reality in some cases. The goal of loosely timed simulation is (i) to make it possible to test functionality that depends on timing, e.g. software using timers, and (ii) to obtain rough performance estimates under explicit assumptions. For example, one may obtain performance estimates of an application assuming that the cache will suffer a known average cache miss. The timing obtained might be wrong, but under correct assumptions, loosely timed simulation are very fast and quite useful.
- *Timeless simulation* is at the highest end of the spectrum. The system does not maintain any timing and the tasks are organized in large functional blocks. Such simulations can be significantly accelerated as the simulation tasks can be highly abstracted. The software can be run and the hardware models can be functionally validated but no performance estimate can be obtained. Such models are also called *programmers view*, as they reflect the hardware behavior as it is perceived by the software programmers.

Timeless models are very useful to do early-prototyping of new software or hardware functionality. Using loosely timed models becomes necessary to validate embedded software and verify basic time-related properties, such as checking that the proper behavior happens after a time-out. But it can hardly be used to obtain performance estimates. As many systems require some performance estimate of the system under design, the community had to address the issue of providing performance estimation of the simulated system. Several directions have been explored that we will discuss in the context of computation and communication models in subsequent sections.

1.4 Computation Abstraction

To illustrate computation abstraction, let us consider the naive example of an integer multiplication. A hardware multiplier component is implemented with multiple additions and shifts. Simulating the multiplication of two numbers A and B by replicating

Fig. 1.3 Interpretive
simulation

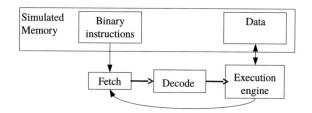

the shift and adds in software would obviously be slow. Comparatively, using the C
code $result = A * B$; is more efficient. It is a typical computation abstraction: to
obtain the result of a computation, a different technique is used from the one used
in the real system, but one trusts (and hopefully one can prove) that the result of the
two computations are identical. In doing the multiplication using C, one trusts that
the host computer carrying out the operation will indeed yield the same result, but
much faster than replicating the shifts and adds.

As a lot of a simulation session is spent in executing processor instructions, *processor simulation* is a critical item of a virtual prototyping platform. Each processor
implements a specific instruction set, related to a particular computer architecture.
Hence the term instruction set simulator (ISS) is used, which may cover several actual
processors with minor variants. ISSs are a common form of computation abstraction.

1.4.1   Instruction Set Simulation

An instruction set simulator is used to mimic the behavior of a target computer processor on a simulation host machine. Its main task is to carry out the computations that
correspond to each instruction of the simulated program. There are several alternatives to achieve such simulation. In *interpretive simulation*, each binary instruction
of the target program is fetched from memory, decoded, and executed, as shown in
Fig. 1.3. This method is flexible and easy to implement, but the simulation speed
is relatively slow as it wastes a lot of time in decoding. Interpretive simulation has
been used in tools such as Insulin [80], Simplescalar [7] and various architecture
exploration tools [29].

A second technique called *static translation* or *compiled ISS* is based on a direct
translation of target instructions to be simulated into one or more instructions in the
host's instruction set. These native instructions manipulate a model of the emulated
target machine state in a functionally equivalent manner. In statically compiled ISSs,
this translation is performed once for the complete target binary before the simulation
is started. The simulation speed with static translation is vastly improved [57, 60, 70,
80, 87, 89], but such approaches are less flexible and do not support self-modifying
or dynamically loaded code.

A third technique to implement an ISS is *dynamic binary translation*. It has
been pioneered by [13, 83], with ideas dating back to dynamic compilation used

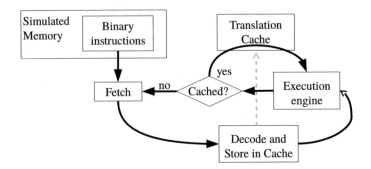

Fig. 1.4 Dynamic binary translation

in the earliest virtual machine based languages [15]. With dynamic translation, shown in Fig. 1.4, the target instructions are fetched from memory at runtime, like in interpretive simulation. But they are decoded only on the first execution and the simulator translates them into another representation, which is stored into a cache. On further execution of the same instructions, the translated cached version is used and its execution will be faster. If the target code is modified during runtime (for example by self-modifying programs) the simulator must invalidate the cached representation.

Dynamic translation adds some translation delay to the total simulation time (unlike static translation) but the translation pay-off is so large that it is usually worthwhile. Although it is not as fast as static translation, dynamic binary translation supports the simulation of programs for which one does not have the complete source code, one does not have the auxiliary libraries available on the host, or the application does dynamic loading or uses self-modifying code. Hence, many ISSs nowadays are dynamic translation simulators [2, 30, 36, 53, 60, 63, 71, 76, 77] with a reasonable speed.

Dynamic cached translators all work on the same model, but they can be subdivided into roughly two categories according to the nature of the translated data and its usage, which we will call here *object-oriented* or *native code generators*.

In an *object-oriented* ISS, the original binary instructions are each translated into an object, in the sense of object-oriented languages, which captures the variable data from the instruction and defines methods that are called upon execution. Object-oriented ISSs are not so difficult to construct. In fact, using some appropriate input formalism, the code of the object classes and methods can even be generated. Using this technique, the ISS is independent of the host operating system and independent of the host processor, therefore easily portable. Because the methods can be compiled in advance, a compiler with maximum optimizations can be used. When constructing an instruction object, the constructor initializes its variables with the instruction data for the methods to operate. Object oriented ISSs can reach speeds above 100 MIPS on a standard computer.

An optimization that can be used in dynamic translation consists of analyzing the binary at decoding time to reconstruct the control flow graph of the application

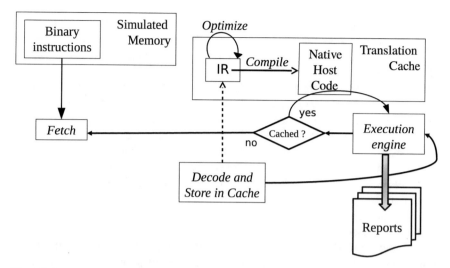

Fig. 1.5 Dynamic binary translation to native code with IL

software in basic blocks, each ending with a branch instruction. Although it may happen that the destination of the branch is unknown at decoding time, the edges of the graph can be dynamically constructed as the simulation advances. The dynamic translation can translate basic blocks into a single object and achieve further optimizations. Another possible optimization is the usage of a partial evaluation compiling technique [19] in which a number of parameters for simulating each instruction are discovered at decoding time and specialized versions of the instructions can be used instead of the generic one [30]. In general, the target code to be translated, whether statically or dynamically, can be decomposed into *translation units*, i.e., sequences of instructions that have a logical structure, e.g. nodes from the control flow graph. Each translation unit is executed individually in the simulation. Upon exit of a unit, control returns to the main simulation loop, which makes it possible to handle various operations such as checking for interrupts or gathering statistics. Increasing the size of translation units usually makes it possible to achieve further optimizations and increase simulation speed, but also decreases accuracy, as discussed below with respect to interrupts.

With the *code generation* technique, the ISS dynamically translates the target binary code into host native code. There are again variants in this technique: (i) re-generate C code that can be dynamically recompiled [63], or (ii) generate some intermediate language code for which an existing back-end compiler is used to generate the host code [31, 77] as illustrated in Fig. 1.5. Technique (i) is easier to implement but adds a significant translation delay to recompile C source code. Technique (ii) can take advantage of existing compiler intermediate languages and back-ends, such as LLVM [41] or GCC GIMPLE [10] intermediate representations.

A variant of that is the *micro-instruction* technique, initially used in QEMU [2]. Each instruction is translated into a sequence of pre-defined micro-instructions that are stored into the cache. These pre-defined micro-instructions can be considered as a very low-level virtual machine that is run inside the ISS, but this virtual machine is strongly related to the way the simulated processor state is represented inside the ISS. In order to translate each instruction, the parameters of each micro-instruction must be extracted, and then the micro-instructions from an entire basic block can be "glued" together into a sequence. Because a whole sequence of target instructions can be translated into a block of micro-instructions that directly modify the virtual state, the execution is faster and a higher simulation speed can be reached. The micro-instruction code be compiled in advance with a highly optimizing compiler and pre-loaded in memory. But the micro-instructions must undergo a linkage process that may be dependent upon both target binary format and host operating system, hence reducing portability of the ISS.

The code generation technique provides higher performance than the object-oriented technique. The Edinburgh University ISS claims to reach performance in the range of over 400 MIPS [36]. The SocLib project reports [25] speedups with a factor of several hundreds in a comparison between their virtual prototyping framework and their cycle-accurate simulator. However, this performance must be balanced with the throughput. Because dynamic translation is more complex, it takes more time and the simulation time must be long enough to amortize the translation time. Dynamic translation with native code generation can become comparable in speed to statically compiled or even natively executed code, allowing complete operating systems to be booted in a virtual prototype.

In naive translation, each instruction is executed as a function in native code. Moving up at the block level, using a compiler intermediate language such as LLVM, the functions simulating each of the instructions from a block can be grouped into one function. The compiler optimizer can be called to do optimizations such as in-lining and constant propagation on that function, which is then compiled into native code. In the end, the size of the native translated block is much smaller than the sum of the individual functions and it executes several instructions at a much higher speed [5, 35]

Interpreted, static or dynamic translation approaches are interesting for simulating platforms based on existing commercial off-the-shelf processors, when one can leverage an existing tool. But there are also cases where one wants to explore a new architecture for which there are no available simulators. As it is a large effort to build a simulator from scratch, it is advantageous in such cases to generate the simulator from a higher-level architecture description language (ADL). These languages have been designed specifically for generating either compilers or simulators, and have been classified [50] into three categories:

- *Structural* models such as MIMOLA [88] that focus on the detail of the micro-architecture, to explore the low-level architecture design. Such approaches are in fact oriented towards circuit synthesis rather than virtual prototyping, although MIMOLA has been used to generate simulators [45].
- *Behavioral* languages such as ISDL [26] or nML [18] that focus more on the instruction sets and their simulators (or compilers).

- *Mixed* ones that allow both, such as LISA [58], EXPRESSION [51], MADL [65], or Harmless [37].

Both behavioral or mixed languages can be used to generate simulators that can be completed and instrumented to build an exploratory virtual prototype.

In order to provide timing information in the context of instruction-set simulation, the ISS needs to be augmented with an instruction-level timing model. In interpreted ISSs, this can be a detailed, cycle-by-cycle model of how different instructions propagate through the processor-internal pipeline, including their interactions with and through the micro-architectural machine state in every cycle. Such detailed micro-architecture models can provide cycle-accurate results, but are expensive to simulate. At the other end of the spectrum, when speed is the primary concern, one can simply associate a fixed number of cycles with each instruction, e.g. either assuming a constant cycles per instruction (CPI) or using instruction-specific table lookups. Such approaches incur little to no simulation overhead, but are unable to capture dynamic timing behavior in modern deeply pipelined, out-of-order processor architectures.

Several alternative approaches have been explored to provide fast yet accurate ISSs. For one, computer programs exhibit repeating behavior. To estimate performance it may not be necessary to run all of the program using accurate timed simulation. The performance measured on a sequence of instructions will be the same when this sequence of instructions is repeated in the same conditions. The *sampling* simulation technique is achieving precisely that: given properly selected samples of instructions, only these instructions are simulated with accurate timing. The remaining instructions can be executed using timeless simulation. Sampling can draw upon statistics theory to estimate a property of a population by examining only a subset of this population. Of course, the difficulty is in choosing the appropriate samples that represent the repeated behavior.

Sampling benefits simulation speeds, which are order of magnitude faster (from 30 to 60 times faster) than with cycle-accurate simulators. Known representatives of sampling methods are Simpoint [27, 59], SimFlex [28], SMARTS [84] and EXPERT [46]. These systems differ mostly in the manner the samples are selected in size and frequency. Sampling techniques can provide fairly accurate performance prediction with a high probability, but may also generate very large data files. A difficulty with sampling is also the transition from timeless to timed models. When a series of instructions have been run in functional simulation, the details of architecture state are usually lost, but the architecture state must be initialized properly to start running in timed mode and obtain accurate estimates. This problem is known as the *warm-up* issue. Before starting timed simulation, the details of hardware state must be reinitialized. This can be done by saving the results from previous simulations or computing an estimated state. However, establishing a trustworthy state may require to roll-back into long program history. The warm-up phase introduces uncertainty in the results and simulation slow down. Given appropriate program sampling, existing technologies report that they can estimate performance within a 3% error margin with a confidence larger than 99.5%.

Another approach consists in statistical workload generation and simulation after collecting data from one complete simulation [54, 66]. Because cache misses are the major performance bottleneck, specific attention has been devoted to cache performance and cache behavior prediction. For example, statistical methods have been successfully used for cache analysis in [3]. Yet another approach consists in compile time static analysis of the code for predicting cache behavior [81] based on cache miss equations. This method is very fast, but limited in the scope of programs that can be analyzed, and it requires access to the source program.

1.4.2   Source-Level and Host-Compiled Simulation

As an alternative to ISSs, so-called *source-level simulation* has recently emerged as an approach that is based on translation of the application source code to be simulated into another program compiled for the host machine. This technique does not use an ISS and is not a straightforward compilation. As such, it avoids the overhead of having to emulate a non-native instruction set. Instead, the translator knows the target system and it directly generates code specifically optimized for it. In simple cases [4, 9], the application source code is translated into C code similar to the original one, with only additions or modifications to handle non-portable aspects in the source code, such as translation of data types into target-equivalent bit widths. In more sophisticated translations [11, 32, 79, 82], as illustrated on Fig. 1.6, the source code is first translated into a lower-level representation similar to a compiler intermediate representation, which can be further analyzed. If the intermediate representation includes the control flow graph, the graph can be analyzed and function calls may be added at each transition. Finally a new source code program is generated after the transformations, which is compiled for the host computer and linked with the simulation runtime and libraries.

So-called *host-compiled* approaches extend source-level simulation of application code with models that emulate the complete software execution environment [6, 22]. Pure source-level approaches can not simulate binary-only library or operating system code that is closely tied to the target processor hardware and instruction set. Instead, any such code needs to be explicitly emulated in the virtual prototype. This requires a specific runtime component to be linked with the translated code. In the simplest case, the translator has some objectives of quantifying or qualifying some measurements and properties. It augments the original source code with function calls to the runtime component so that the simulator can perform internal actions, e.g. to collect statistics, tracing memory usage, check assertions, etc.

In particular, however, operating system calls may be detected and transformed into calls to a *virtual operating system* or an *operating system model*. For example, if the source code contains a call to `openfile()` on the target OS, this call may be detected and replaced with a call to the virtual OS model. This call can then decide to instead open a local file on the host operating system with some mapping information provided as a simulation parameter, e.g. to adapt to a different directory structure.

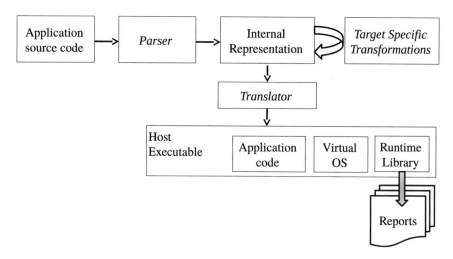

Fig. 1.6 Host-compiled simulation

A similar *semi-virtualization* of OS calls into equivalent host functionality is often also applied in ISSs.

A key aspect of OS models in host-compiled approaches is also to accurately emulate the multi-tasking behavior beyond just a single source-level task as it would occur in reality on the simulated target system. A host-compiled simulator can not run the binary code of a real OS to manage the interactions between multiple tasks. One approach is to instead port and para-virtualize the target operating system to execute on top of the host's OS kernel. However, this requires OS sources to be available and often incurs a significant porting effort and simulation overhead. Alternatively, lightweight OS models can be developed to provide a simulation of task scheduling, communication and synchronization behavior directly on top of languages like SystemC [23, 40, 42, 61]. The virtual OS model then must be a part of the simulation runtime component on the host machine. Similarly, the function calls into auxiliary target libraries and drivers for interaction with external peripherals may then also be detected and transformed into calls into a host library, OS model or the overall system and communication simulation environment.

A key concern in source-level and host-compiled models is accurate timing simulation at such a high level of abstraction. A technique that has become popular and that can be applied for both the host-compiled approach and dynamic ISS translation is the technique of *back-annotation* [11]. The main idea is to analyze the structure of the simulated code at some granularity. The analysis is related to some objective and model, for example to estimate performance, and results into some estimates for certain metrics. The data obtained with this analysis can then be used to back-annotate the code, as kind of meta-information, for further re-use during the simulation. For the sake of generality and to analyze virtual prototypes of applications that can be ported to multiple platforms, the code to annotate can be analyzed at some intermediate, target-independent level. However, the target compiler might still optimize the code with techniques such as loop unrolling and inlining, making the generic estimates over-pessimistic. Although it is more costly, a more accurate analysis can be done by using the target-dependent basic block structure effectively generated by the target compiler's IR. In either case, the technique consists of annotating the basic blocks with additional performance or power consumption data that is later used to compute non-functional properties during simulation.

Let us illustrate this concept of back-annotation with an example drawn for performance estimation of an e200 processor from the Power architecture family. This processor has a 2-issue pipeline. Thus, in general it can simultaneously execute 2 instructions. The pipeline takes its instructions from an instruction buffer of 8 instructions that is itself refilled regularly from the instruction cache. Considering a basic block, one can do the following to estimate the number of cycles used by the block execution. First, knowing the instruction cache contents at the entrance of the block, one can compute whether there will be an instruction cache miss during the execution of that block. Since many blocks are smaller than the cache lines held, there will be relatively frequent cache hits, simplifying the block analysis. If there is a cache miss, it is possible to predict at which instruction(s) the cache miss(es) will occur, what consequently the new contents of the cache will be, and to compute the delay, if any. A similar approach can be applied to model data caches and associated hit/miss penalities on every load/store instruction [48, 56, 62]. However, accurately tracking cache state to determine hits and misses at the source level requires relatively complex analysis to reconstruct physical address traces that in turn drive a dynamic and thus slow to simulate behavioral cache model.

Another potential source of delays are the pipeline hazards when two instructions in each issue of the pipeline want to access the same resource. For example, if the same register would be accessed at the same time or two simultaneous load/store instructions would occur, but the e200 only has one load/store unit. These hazards can be detected by analyzing the block, dispatching of instructions into the pipeline according to an abstract e200 pipeline model, and calculating the resulting delays. For hazards that span across block boundaries, models can track the actual control flow paths taken during simulation and adjust delays accordingly. Finally, another source of delay is the branch prediction system. The e200 has a branch unit that includes a branch target buffer (BTB) of 8 addresses. When a branch instruction is entering the pipeline, it checks whether it is already in the BTB, and possibly refills the BTB. Because branch instructions may only occur as the last instructions of the basic blocks, the branch analysis can be done only during the transitions between blocks [17].

As seen on this example, some of the performance analysis made at a block level can be done statically only once, and some dynamically, but only at the entrance and exit of a block. Hence, the block can be annotated with the results of such analysis either with purely static data, and/or with dynamic hooks to perform updates and compute additional delays. Eventually, the whole simulation will gather the static and dynamic data to obtain performance estimation.

This type of analysis and back annotation can clearly be done within an ISS using dynamic binary translation, but also within a host-compiled approach that analyzes the code generated by the target compiler, obtains the data, and back-annotates the source or IR code of the application before compiling [4, 11, 32, 79, 82]. Not only is the simulation quite faster, but the back-annotation mechanism makes it possible to check other non-functional properties of the application [86]. Some people also now combine host-compiled approaches with dynamic translation [20].

Note that the simulation speed and accuracy can be affected significantly by the level of abstraction defined by the virtual OS. The virtual OS can be used to map simple system calls that mimic the target OS behavior, such as managing files or obtaining a network connection, but it can also be used to study various operating system scheduling policies or predict non-functional aspects of the software. An issue arises when temporal decoupling or time warping abstractions are applied to speed up simulations. Asynchronous task events may be captured at wrong times leading to inaccuracies in the simulated scheduling order at coarser timing granularities. To overcome speed and accuracy tradeoffs, both conservative [67] and optimistic [74] OS simulation approaches have been developed. Similarly, simulating the correct order of accesses to shared resources such as caches in a multi-core context requires corresponding solutions for fast yet accurate modeling of cache and resource contentions [68, 78]. Finally, host-compiled simulations can be augmented with models of other processor and hardware behavior, such as exceptions, interrupt handling and external bus communication [21, 38, 69, 75] (see also Sect. 1.5).

The host-compiled approach is the technique used for the iPhone simulator cited above, using a virtual iOS. A drawback is that it is not as flexible as interpretive simulation. The application program source code must be available at translation time, before simulation starts, which is not always the case. Furthermore, one must have both the dependent libraries and the virtual OS available. Host-compiled translation is clearly not suitable for application programs that dynamically modify the code or dynamically load new code at runtime. It does not allow for simulation of virtual machines that generate code on the fly with JIT compilers, not for real operating systems. It can be very useful, however, for testing embedded software applications.

1.5   Communication Abstraction

Hardware components communicate together, and abstracting the communication among them may provide a significant speedup in simulation. In the physical circuit, components are connected together using wires or complex buses. The details of the signals sent for communication are often irrelevant to the virtual prototyping goals. The communication details may be abstracted in the simulation with a higher level message passing schema. The term transaction-level modeling (TLM) has been forged to name the modeling style whereby communication among components is achieved through a uniform *transaction interface* [16, 24] at a level of function calls above pins and wires. This interface hides and abstract the details of the operations.

1.5.1 Transaction-Level Modeling

Transaction-level modeling is a mature and widely adopted topic with a rich literature of existing reference materials [24]. As such, we only briefly summarize key abstraction concepts and ideas here. In a transaction-level model, the component starting the communication is named the *initiator*, it sends a transaction, a message specifying some operation to be performed, to a destination component named the *target*. Both the initiator and the target do not need to be aware of the interconnection structure. It may be a direct transfer or it may be a complex bus structure. TLM makes it possible to implement multiple kinds of models. Purely functional models can use transactions that use highly abstracted interconnect and ignore all communication issues. However, the interface must be standardized if one wants to exchange models among third parties. Because it accommodates various levels of modeling, the open source TLM library from the Open SystemC Initiative (OSCI) [55] has become widely used in academia and industry as it offers an interoperable interface with a fast implementation. In that implementation, a transaction is basically a function call from the initiator, issued on some communication port, which, after routing, is implemented as a function call in the target module.

TLM is not totally independent of the time abstraction, which raises an issue. Indeed, if one wants to build an approximately timed virtual prototype, when issuing transactions to remote components, one wants to know the approximate duration of that transaction and the transaction interface must support that. Virtually all hardware simulation techniques have similar requirements, but we illustrate the problem here in SystemC with the TLM library. The simulation kernel maintains a clock to measure elapsed time. The clock advances through an interface, in SystemC, the `wait()` call. But, whereas the initiator wants to control the timing model of an operation that may require issuing multiple transactions in parallel, if the targets make the clock move up unknowingly of the initiator, it is problematic.

Thus, the TLM standard has developed two models named the *blocking model* and the *non-blocking model*. In the blocking model, transactions are allowed to call `wait()` and the clock may advance during the transaction. It is simpler, and it is possible to build loosely timed models or timeless models that simply ignore the clock value.

In the non-blocking model, transactions must not make the clock move up, but they must provide in the end a timing estimate of the operation. An issue is that the target may be in the situation where it cannot achieve the transaction immediately, or it has no clue at that moment on how long it will take. Hence, the OSCI standardization committee decided to split such transactions into phases, with the so-called *4 phases protocol*. In the first phase, the initiator issues a **request**. The target then replies with an **end-request** to either acknowledge the transaction or raise an error. At this point the initiator may issue other transactions if necessary, but at some point it must wait (with or without moving the clock) for the target response. When the target has done its work, it issues a backward **begin-response**, to which the initiator should reply with an **end-response**. Finally, the transaction ends with a terminating status code.

On each phase, the initiator receives a delay parameter. This delay combines all the delays from the target and the interconnect, e.g. bus contention delays can be added to the target delays. Eventually, the initiator can combine all of the successive or parallel delays recorded during the transactions in the appropriate manner to update the clock with the accurate number of cycles.

In conclusion, loosely timed models can abstract the interconnect to the desired level using blocking transactions. Approximately timed models should implement interconnects descriptions that yield accurate estimates of the communication time and use non-blocking transactions. Sometimes, it is also desirable to mix loosely timed and approximately timed simulation models as long as there is interoperability between them. This can be useful, for example, to evaluate performance of an application where already existing components, whose performance has been evaluated in the past, are combined with a new specific co-processor under design, for which one wants to have precise measurements. Mixing models is a demand, and it is possible to implement it with the TLM library.

Virtual prototype designers can choose the level of abstraction they want for implementing communication abstractions. Although the complexity increases with the 4-phase protocol, data can still be transferred in chunks using software function calls with reasonable performance. To achieve even higher abstraction and speed, temporal decoupling and time warping techniques can be applied on top of blocking (or non-blocking) transfers. The TLM library includes a so-called *quantum keeper* for this purpose. The quantum keeper maintains a configurable time quantum by which each simulated component is allowed to run ahead of the current simulator time. Without further adjustments, however, this leads to transactions being simulated out-of-order, where the time quantum can be adjusted to navigate the associated speed and accuracy tradeoff. Alternatively, various conservative [47, 78] and optimistic [64, 73] TLM simulation approaches have been investigated to maintain or restore the correct transaction order in accesses to shared resources, such as busses, peripherals or the memory system, for which order matters.

1.5.2   Memory Simulation

Memory load and store instructions are the most frequently used transactions in the communication between a CPU and the memory system. Consequently, simulation of memory accesses rapidly becomes the bottleneck. There are two critical points in the simulation of memory accesses: the memory read/write transactions and the simulation of the hardware Memory Management Unit (MMU).

In addition, virtually all architectures use memory-mapped I/O, meaning that communication with hardware peripherals is achieved through load/store instructions. A virtual prototype may adopt a simplistic approach to memory simulation, by only differentiating which memory areas are mapped to peripherals and which are mapped on real memory. For the peripherals, memory operations can be transformed into communications transactions, and the real memory accesses can be mapped onto an equivalent host memory space.

However this simplistic approach may turn out to be inadequate for checking memory protection and performance issues. The various types of real memory used in the target embedded system platform (i.e. static or dynamic memory, ROM, flash memory) are simulated using host memory, but they have different performance characteristics. In a transaction-based simulation, memory accesses would normally each require a transaction, with computation of access time for an approximately timed estimate. However, for a loosely timed simulation, one wants to use host memory to simulate target memory as fast as possible. The TLM 2.0 Direct Memory Interface [55] specification offers an interface so that a memory target segment can reliably be implemented using a dedicated host memory segment and a target memory access implemented at roughly the speed of a host memory access. The host memory and the target memory may also be of different endianness and of different word size. In general, faster simulation is achieved by using the host endianness and word size to represent target memory, and carry out conversions only when needed.

The MMU is the hardware component that controls each memory access and enforces the policy set by the software. On each memory access, it checks whether the memory access is authorized. Otherwise, it routes the instruction to an exception mechanism. Additionally, on some architectures such as ARM and Power, the MMU performs partly or totally the translation of logical addresses to physical addresses. A full system simulator must hence also simulate the MMU, and this becomes a critical element of the overall performance [49, 72]. Modern processors use the notion of logical memory. Typically, the memory logical space is divided up into sections or pages of some size, and a logical page is mapped (or not) into a real page of the physical memory. The operating system must maintain tables associating logical IDs with physical memory addresses. The first task of the MMU is to check whether or not a logical address is mapped or not. It does this by using an associative table named the translation lookaside buffer (TLB) associating within each entry the logical to the physical memory mapping and its access permission rights.

If the TLB lookup fails because there is no mapping, on some architectures an exception is raised named a *TLB miss* or a *page fault*. When a page fault occurs, the operating system must either cancel the faulty program or fill the TLB with appropriate data from the page table. On more sophisticated architectures, the hardware directly performs a *page table walk*. In that case, the MMU also has access to the page table. It searches for the requested entry and inserts it automatically into the TLB. If that table walk also fails, then an exception is raised, allowing the software to terminate the program or provide a new page table. While associative hardware search performance increases with a larger TLB size, the simulator performance decreases. All of these steps can be implemented with engineering tricks to accelerate the process, but memory access simulation remains a performance bottleneck.

Another aspect of memory management, when the simulation is using dynamic translation, is the coherency between the binary target code and the translation cache. Initially, instructions from memory locations are translated into the translation cache. However, this code may be replaced with other code, typically when the operating system is loading a new program in memory to replace a terminated program. The translation cache then becomes inconsistent with memory contents. The simulator must consequently detect memory write operations that overwrite previously translated code to invalidate the corresponding cached translation. Garbage collection of the translation cache must occur with replacement of the initial instructions with new code.

1.5.3 Interrupts

All simulators, whichever technology they use, must handle hardware interrupts, if they want to fully simulate the processor and run operating systems with interrupt handlers. An ISS can mimic the processor behavior by checking for interrupts after every instruction. Doing this by software becomes a performance burden. But one can note that in most systems, the fact that an interrupt occurs exactly at one point in the code should not matter. In fact, an operating system that would be sensitive to the moment when interrupts do occur would be totally unreliable. An ISS can thus increase simulation speed by checking for interrupts at larger intervals, avoiding many useless tests.

Checking for interrupts in the simulation loop is easier with interpretive simulators, for example every N instructions. Dynamic binary translation simulators must plan for checking interrupts at appropriate intervals inserted into the code as simulating long translation units may lead into issues with time related software. Similarly, host-compiled simulators must incorporate appropriate models of the interrupt handling chain that tightly integrate with virtual OS models and emulation of scheduling and processor suspension behavior [69, 85].

1.5.4 Peripherals

A virtual prototype typically includes one or more ISSs (for each core) connected to co-processors and peripherals. It is often the case that the system under design includes some application-specific hardware, for example digital-to-analog converters, data compressing or decompressing components, cryptographic components, etc. If the focus of the design is such a component, the simulation model for this component becomes the focus of the virtual prototype. In early exploratory phases, high-level, purely functional models can be used (e.g. of a crypto algorithm) to abstract the computation. In later phases where more accurate timing behavior of the hardware component and overall system is desired, functional models can be augmented with timing information, e.g. using similar back-annotation based approaches as in source-level and host-compiled software models [43, 44]. Conversely, it may be that the complete application program is not necessary to validate the peripheral component. The ISS can then be replaced by a simple testbench that emits commands to the co-processor.

1.6   Summary and Conclusions

In the development of embedded computer systems, architects want to explore hardware alternatives, software engineers want to develop and validate the software before the hardware platform is available, and both want to verify properties of the system. Virtual prototyping makes this possible. However, there is no virtual prototyping solution that would be as fast and as accurate as the real platform that it virtualizes. Tradeoffs have to be made to obtain virtual prototypes that satisfy the designer's objectives and allow for reasonably paced development iteration cycles. These tradeoffs are obtained by abstracting some aspect of the real prototype into a higher-level model. Designers can choose to explore the modeling space by raising the abstraction level in one of two possible directions, computation and communication, where each can be abstracted in terms of functionality or non-functional properties, chiefly among which is time. Raising the level of abstraction unfortunately results in loosing some accuracy: the designers must choose the most appropriate tradeoff with respect to the virtual prototyping goals.

In fact, a virtual prototype of a system may evolve in time. It may move from a purely functional model at the beginning of a project towards an approximately timed model when reaching the optimization stage. We have presented in this chapter the directions to abstract hardware functions into simulation models by abstracting functionality and time for computation and communication. The various abstractions that can be made and the technologies that can implement such tradeoffs have been presented; and some techniques and standards that make it possible to gather individual models into complex platforms have been mentioned. Virtual prototyping is now a mature enough technology to be used in industrial processes.

A new challenge is the validation of large heterogeneous, fault-tolerant cyber-physical systems (CPS) that have to operate over a network of sensors, actuators and dedicated functions operating in a distributed manner. Embedded computer systems and their virtual prototypes are an integral part of any CPS, but solutions for holistic prototyping of complete CPS do not yet exist, or just recently have started to appear and be developed [14, 52]. A complete virtual CPS prototype may need to be modeled partly with languages such as SysML for the high-level application model, partly with SystemC or similar approaches for the hardware/software components, partly with MatLab/Simulink or Modelica if there is a continuous process in the application and the simulation of that process requires solving equations, and partly in a network simulator to model physical layers and protocol stacks of device-to-device communications. In order to build a virtual prototype of such a large system, it is necessary to interconnect continuous simulators based on solving differential equations with discrete-event simulators and general modelers. To test various properties of the system, in particular to verify that the overall application software is fault-tolerant to potential hardware failures of the components and satisfies the imposed constraints in the continuous process, one needs to control the overall execution of the individual simulators with synchronization steps and data exchange protocols, and this is not yet state-of-the-art.

Finally, as CPS have become more widely used in many applications that are safety-critical for human beings, it becomes even more necessary to guarantee a proper system behavior. We believe that virtual prototyping frameworks should be more strongly coupled with formal verification tools such as model checkers, error finding methods such as abstract interpretation or trace analysis, and verification of non-functional properties, possibly using theorem provers, to correlate the simulation results with formal requirements and finally achieve faster validation of systems with stronger proofs of correctness.

References

1. G. Arnout, SystemC standard, in *Proceedings of the Asia South Pacific Design Automation Conference, ASPDAC* (2000), pp. 573–578
2. F. Bellard, QEMU, a fast and portable dynamic translator, in *Proceedings of the USENIX Annual Technical Conference, ATEC* (2005), p. 41
3. E. Berg, H. Zeffer, E. Hagersten, A statistical multiprocessor cache model, in *Proceedings of the IEEE International Symposium on Performance Analysis of Systems and Software, ISPASS* (2006), pp. 89–99
4. A. Bouchhima, P. Gerin, F. Pétrot, Automatic instrumentation of embedded software for high level hardware/software co-simulation. in *Proceedings of the Asia and South Pacific Design Automation Conference, ASPDAC* (2009), pp. 546–551
5. F. Brandner, A. Fellnhofer, A. Krall, D. Riegler, Fast and accurate simulation using the LLVM compiler framework, in *Methods and Tools, Rapid Simulation and Performance Evaluation, RAPIDO* (2009)
6. O. Bringmann, W. Ecker, A. Gerstlauer, A. Goyal, D. Mueller-Gritschneder, P. Sasidharan, S. Sing, The next generation of virtual prototyping: ulta-fast yet accurate simulation of HW/SW systems, in *Proceedings of the Design, Automation and Test in Europe Conference, DATE* (2015)
7. D. Burger, T. Austin, D. Burger, T.M. Austin, The simplescalar tool set, version 2.0. Technical Report TR-1342 (University of Wisconsin-Madison, 1997)
8. L. Cai D. Gajski, Transaction level modeling: an overview. in *Proceedings of the 1st IEEE/ACM/IFIP International Conference on Hardware/software Codesign and System Synthesis, CODES+ISSS* (2003), pp. 19–24
9. L. Cai, A. Gerstlauer, D. Gajski, Retargetable profiling for rapid, early system-level design space exploration. in *Proceedings of the 41st Design Automation Conference, DAC* (2004), pp. 281–286
10. S. Callanan, D.J. Dean, E. Zadok, Extending GCC with modular GIMPLE optimizations, in *Proceedings of the 2007 GCC Developers Summit* (2007), pp. 31–37
11. S. Chakravarty, Z. Zhao, A. Gerstlauer, Automated, retargetable back-annotation for host compiled performance and power modeling, in *Proceedings of the IEEE/ACM/IFIP International Conference on Hardware/Software Codesign and System Synthesis, CODES+ISSS* (2013)
12. D. Chiou, D. Sunwoo, J. Kim, N. Patil, W.H. Reinhart, D.E. Johnson, Z. Xu, The FAST methodology for high-speed SoC/computer simulation, in *Proceedings of the 2007 IEEE/ACM International Conference on Computer-Aided Design, ICCAD* (2007), pp. 295–302
13. B. Cmelik, D. Keppel, Shade: a fast instruction-set simulator for execution profiling, in *Proceedings of the 1994 ACM Conference on Measurement and Modeling of Computer Systems, SIGMETRICS* (May 1994), pp. 128–137
14. DESTEC Consortium. Design support and tooling for embedded control software (DESTECS), http://www.destecs.org/

15. L.P. Deutsch, A.M. Schiffman, Efficient implementation of the Smalltalk-80 system. in *Proceedings of the 11th ACM SIGACT-SIGPLAN Symposium on Principles of Programming Languages, POPL* (1984), pp. 297–302

16. A. Donlin, Transaction level modeling: flows and use models, in *Proceedings of the 2nd IEEE/ACM/IFIP International Conference on Hardware/Software Codesign and System Synthesis, CODES+ISSS* (2004), pp. 75–80

17. A. Faravelon, N. Fournel, F. Ptrot, Fast and accurate branch predictor simulation, in *Proceedings of the Design Automation and Test in Europe Conference, DATE* (2015), pp. 317–320

18. A. Fauth, J. Van Praet, M. Freericks, Describing instruction set processors using nML, in *Proceedings of the 1995 European Conference on Design and Test, EDTC* (1995), p. 503

19. Y. Futamura, Partial evaluation of computation process-an approach to a compiler-compiler. High. Order Symbolic Comput. **12**(4), 381–391 (1999)

20. L. Gao, K. Karuri, S. Kraemer, R. Leupers, G. Ascheid, H. Meyr, Multiprocessor performance estimation using hybrid simulation, in *Proceedings of the Design Automation Conference, DAC* (2008), pp. 325–330

21. P. Gerin, H. Shen, A. Chureau, A. Bouchhima, A. Jerraya, Flexible and executable hardware/software interface modeling for multiprocessor SoC design using SystemC, in *Proceedings of the Asia and South Pacific Design Automation Conference, ASPDAC* (2007)

22. A. Gerstlauer, S. Chakravarty, M. Kathuria, P. Razaghi, Abstract system-level models for early performance and power exploration, in *Proceedings of the Asia and South Pacific Design Automation Conference, ASPDAC* (2012), pp. 213–218

23. A. Gerstlauer, H. Yu, D. Gajski, RTOS modeling for system level design, in *Proceedings of the Design, Automation and Test in Europe Conference, DATE* (2003), pp. 130–135

24. F. Ghenassia (ed.), *Transaction-Level Modeling with SystemC. TLM Concepts and Applications for Embedded Systems* (Springer, New York, 2005). ISBN 0-387-26232-6

25. M. Gligor, N. Fournel, F. Pétrot, Using binary translation in event driven simulation for fast and flexible MPSoC simulation, in *Proceedings of the IEEE/ACM/IFIP International Conference on Hardware/Software Codesign and System Synthesis, CODES+ISSS* (2009), pp. 71–80

26. G. Hadjiyiannis, S. Hanono, S. Devadas, ISDL: an instruction set description language for retargetability and architecture exploration. Des. Autom. Embed. Syst. **6**(1), 39–69 (2000)

27. G. Hamerly, E. Perelman, B. Calder, How to use SimPoint to pick simulation points. SIGMETRICS Perform. Eval. Rev. **31**(4), 25–30 (2004)

28. N. Hardavellas, S. Somogyi, T.F. Wenisch, R.E. Wunderlich, S. Chen, J. Kim, B. Falsafi, J.C. Hoe, A.G. Nowatzyk, SimFlex: a fast, accurate, flexible full-system simulation framework for performance evaluation of server architecture. SIGMETRICS Perform. Eval. Rev. **31**(4), 31–34 (2004)

29. M.R. Hartoog, J.A. Rowson, P.D. Reddy, S. Desai, D.D. Dunlop, E.A. Harcourt, N. Khullar, Generation of software tools from processor descriptions for hardware/software codesign, in *Proceedings of the 34th Annual Design Automation Conference, DAC* (1997), pp. 303–306

30. C. Helmstetter, V. Joloboff, SimSoC: a SystemC TLM integrated ISS for full system simulation, in *Proceedings of the IEEE Asia Pacific Conference on Circuits and Systems, APCCAS* (2008), http://formes.asia/cms/software/simsoc

31. D.-Y. Hong, C.-C. Hsu, P.-C. Yew, J.-J. Wu, W.-C. Hsu, P. Liu, C.-M. Wang, Y.-C. Chung, HQEMU: a multi-threaded and retargetable dynamic binary translator on multicores, in *Proceedings of the Tenth International Symposium on Code Generation and Optimization, CGO* (2012), pp. 104–113

32. Y. Hwang, S. Abdi, D. Gajski, Cycle-approximate retargetable performance estimation at the transaction level, in *Proceedings of the Design, Automation and Test in Europe Conference, DATE* (2008), pp. 3–8

33. IEEE, *Open SystemC Language Reference Manual* (2011), http://standards.ieee.org/getieee/1666/download/1666-2011.pdf

34. D. Jefferson, H. Sowizral, Fast concurrent simulation using the time warp mechanism. Technical report, the Rand Corporation (Santa Monica, California, 1982). Rand Note N-1906AF

35. V. Joloboff, X. Zhou, C. Helmstetter, X. Gao, Fast instruction set simulation using LLVM-based dynamic translation. in *International MultiConference of Engineers and Computer Scientists, IAENG* vol. 2188 (Springer, 2011), pp. 212–216
36. D. Jones, N. Topham, High speed CPU simulation using LTU dynamic binary translation, in *Proceedings of the 4th International Conference on High Performance Embedded Architectures and Compilers, HiPEAC* (2009), pp. 50–64
37. R. Kassem, M. Briday, J.-L. Bchennec, G. Savaton, Y. Trinquet, Harmless, a hardware architecture description language dedicated to real-time embedded system simulation. J. Syst. Archit. **58**(8), 318–337 (2012)
38. T. Kempf, M. Dorper, R. Leupers, G. Ascheid, H. Meyr, T. Kogel, B. Vanthournout, A modular simulation framework for spatial and temporal task mapping onto multi-processor SoC platforms, in *Proceedings of the Design, Automation and Test in Europe Conference, DATE* (2005)
39. K. Keutzer, A. Newton, J. Rabaey, A. Sangiovanni-Vincentelli, System-level design: orthogonalization of concerns and platform-based design. IEEE Trans. Comput. Aided Des. Integr. Circuits Syst. (TCAD) **19**(12), 1523–1543 (2000)
40. M. Krause, D. Englert, O. Bringmann, W. Rosenstiel, Combination of instruction set simulation and abstract RTOS model execution for fast and accurate target software evaluation, in *Proceedings of the 6th IEEE/ACM/IFIP International Conference on Hardware/Software Codesign and System Synthesis, CODES+ISSS* (2008), pp. 143–148
41. C. Lattner, V. Adve, LLVM: a compilation framework for lifelong program analysis & transformation, in *Proceedings of the 2004 International Symposium on Code Generation and Optimization, CGO* (2004)
42. R. Le Moigne, O. Pasquier, J.-P. Calvez, A generic RTOS model for real-time systems simulation with SystemC, in *Proceedings of the Design, Automation and Test in Europe Conference, DATE* (2004)
43. D. Lee, L.K. John, A. Gerstlauer, Dynamic power and performance back-annotation for fast and accurate functional hardware simulation, in *Proceedings of the Design, Automation and Test in Europe Conference, DATE* (2015)
44. D. Lee, T. Kim, K. Han, Y. Hoskote, L.K. John, A. Gerstlauer, Learning-based power modeling of system-level black-box IPs, in *Proceedings of the IEEE/ACM International Conference on Computer-Aided Design, ICCAD* (2015)
45. R. Leupers, J. Elste, B. Landwehr, Generation of interpretive and compiled instruction set simulators, in *Proceedings of the Asia and South Pacific Design Automation Conference, ASPDAC* (1999), pp. 339–342
46. W. Liu, M.C. Huang, EXPERT: expedited simulation exploiting program behavior repetition, in *Proceedings of the 18th Annual International Conference on Supercomputing, ICS* (2004), pp. 126–135
47. K. Lu, D. Muller-Gritschneder, U. Schlichtmann, Analytical timing estimation for temporally decoupled TLMs considering resource conflicts, in *Proceedings of the Design, Automation and Test in Europe Conference, DATE* (2013), pp. 1161–1166
48. K. Lu, D. Müller-Gritschneder, U. Schlichtmann, Fast cache simulation for host-compiled simulation of embedded software, in *Proceedings of the Design, Automation and Test in Europe Conference* (2013), pp. 637–642
49. P. Magnusson, B. Werner, Efficient memory simulation in SimICS, in *Proceedings of the 28th Annual Simulation Symposium* (April 1995), pp. 62–73
50. P. Mishra, N. Dutt, *Processor Description Languages* (Morgan Kaufmann Publishers Inc., San Francisco, 2008)
51. W.S. Mong, J. Zhu, A retargetable micro-architecture simulator, in *Proceedings of the Design Automation Conference, DAC* (2003), p. 752
52. W. Mueller, M. Becker, A. Elfeky, A. DiPasquale, Virtual prototyping of cyber-physical systems, in *Proceedings of the 17th Asia and South Pacific Design Automation Conference, ASPDAC* (2012), pp. 219–226

53. A. Nohl, G. Braun, O. Schliebusch, R. Leupers, H. Meyr, A. Hoffmann, A universal technique for fast and flexible instruction-set architecture simulation, in *Proceedings of the 39th Design Automation Conference, DAC* (2002), pp. 22–27

54. S. Nussbaum, J.E. Smith, Modeling superscalar processors via statistical simulation, in *Proceedings of the International Conference on Parallel Architectures and Compilation Techniques, PACT* (2001), pp. 15–24

55. Open SystemC Initiative. OSCI TLM-2.0 reference manual (2009), http://www.accellera.org/downloads/standards/

56. A. Pedram, D. Craven, A. Gerstlauer, Modeling cache effects at the transaction level, in *Proceedings of the International Embedded Systems Symposium, IESS* (2009)

57. S. Pees, A. Hoffmann, H. Meyr, Retargetable compiled simulation of embedded processors using a machine description language. ACM Trans. Des. Autom. Electron. Syst. (TODAES) **5**(4), 815–834 (2000)

58. S. Pees, A. Hoffmann, V. Zivojnovic, H. Meyr, LISA—machine description language for cycle-accurate models of programmable DSP architectures, in *Proceedings of the 36th Annual ACM/IEEE Design Automation Conference, DAC)* (1999), pp. 933–938

59. E. Perelman, G. Hamerly, M. Van Biesbrouck, T. Sherwood, B. Calder, Using simpoint for accurate and efficient simulation. SIGMETRICS Perform. Eval. Rev. **31**(1), 318–319 (2003)

60. M. Poncino, J. Zhu, DynamoSim: a trace-based dynamically compiled instruction set simulator, in *Proceedings of the 2004 IEEE/ACM International Conference on Computer-Aided Design, ICCAD* (2004), pp. 131–136

61. H. Posadas, J. Adamez, E. Villar, F. Blasco, F. Escuder, RTOS modeling in systemC for real-time embedded SW simulation: a POSIX model. Des. Autom. Embed. Syst. **10**(4), 209–227 (2005)

62. H. Posadas, L. Díaz, E. Villar, Fast data-cache modeling for native co-simulation, in *Proceeding of the Asia and South Pacific Design Automation Conference, ASPDAC* (2011)

63. W. Qin, J. D'Errico, X. Zhu, A multiprocessing approach to accelerate retargetable and portable dynamic-compiled instruction-set simulation, in *Proceedings of the IEEE/ACM/IFIP International Conference on Hardware/Software Codesign and System Synthesis, CODES+ISSS* (2006), pp. 193–198

64. W. Qin, S. Rajagopalan, S. Malik, A formal concurrency model based architecture description language for synthesis of software development tools. SIGPLAN Not. **39**(7), 47–56 (2004)

65. M. Radetzki, R. Khaligh, Accuracy-adaptive simulation of transaction level models, in *Proceedings of the Design, Automation and Test in Europe Conference, DATE* (2008), pp. 788–791

66. R. Rao, M. Oskin, F. Chong, HLSpower: hybrid statistical modeling of the superscalar power-performance design space, in *High Performance Computing*, vol. 2552, HiPC, ed. by S. Sahni, V. Prasanna, U. Shukla (Springer, Berlin Heidelberg, 2002), pp. 620–629

67. P. Razaghi, A. Gerstlauer, Automatic timing granularity adjustment for host-compiled software simulation, in *Proceedings of the Asia and South Pacific Design Automation Conference, ASPDAC* (2012)

68. P. Razaghi, A. Gerstlauer, Multi-core cache hierarchy modeling for host-compiled performance simulation, in *Proceedings of the Electronic System Level Synthesis Conference, ESLSyn* (2013)

69. P. Razaghi, A. Gerstlauer, Host-compiled multi-core system simulation for early real-time performance evaluation, ACM Trans. Embed. Comput. Syst. (TECS) **13**(5s), (2014)

70. M. Reshadi, N. Bansal, P. Mishra, N. Dutt, An efficient retargetable framework for instruction-set simulation, in *Proceedings of the 1st IEEE/ACM/IFIP International Conference on Hardware/Software Codesign and System Synthesis, CODES+ISSS* (2003), pp. 13–18

71. M. Reshadi, P. Mishra, N. Dutt, Instruction set compiled simulation: a technique for fast and flexible instruction set simulation, in *Proceedings of the Design Automation Conference, DAC* (2003), pp. 758–763

72. M. Rosenblum, S. Herrod, E. Witchel, A. Gupta, Complete computer system simulation: the SimOS approach. IEEE Parallel Distrib. Technol. Syst. Appl. **3**(4), 34–43 (1995)

73. G. Schirner, R. Dömer, Result oriented modeling: a novel technique for fast and accurate TLM. IEEE Trans. Comput. Aided Des. Integr. Circuits Syst. (TCAD) **26**(9), 1688–1699 (2007)

74. G. Schirner, R. Dömer, Introducing preemptive scheduling in abstract RTOS models using result oriented modeling, in *Proceedings of the Design, Automation and Test in Europe Conference, DATE* (2008)
75. G. Schirner, A. Gerstlauer, R. Dömer, Fast and accurate processor models for efficient MPSoC design. ACM Trans. Des. Autom. Electron. Syst. (TODAES) **15**(2), 101–1026 (2010)
76. K. Scott, N. Kumar, S. Velusamy, B. Childers, J.W. Davidson, M.L. Soffa, Retargetable and reconfigurable software dynamic translation, in *Proceedings of the International Symposium on Code Generation and Optimization, CGO* (2003)
77. H. Shi, Y. Wang, H. Guan, A. Liang, An intermediate language level optimization framework for dynamic binary translation. SIGPLAN Not. **42**(5), 3–9 (2007)
78. S. Stattelmann, O. Bringmann, W. Rosenstiel, Fast and accurate resource conflict simulation for performance analysis of multi-core systems, in *Design, Automation Test in Europe Conference Exhibition, DATE* (2011)
79. S. Stattelmann, O. Bringmann, W. Rosenstiel, Fast and accurate source-level simulation of software timing considering complex code optimizations, in *Proceedings of the 48th Design Automation Conference, DAC* (2011), pp 486–491
80. S. Sutarwala, P.G. Paulin, Y. Kumar, Insulin: an instruction set simulation environment, in *Proceedings of the 11th IFIP WG10.2 International Conference on Computer Hardware Description Languages and their Applications, CHDL* (1993), pp. 369–376
81. X. Vera, J. Xue, Let's study whole-program cache behaviour analytically, in *Proceedings of the 8th International Symposium on High-Performance Computer Architecture, HPCA* (2002), p. 175
82. Z. Wang, J. Henkel, Accurate source-level simulation of embedded software with respect to compiler optimizations, in *Proceedings of the Design, Automation Test in Europe Conference, DATE* (2012)
83. E. Witchel, M. Rosenblum, Embra: fast and flexible machine simulation, in *Proceedings of the 1996 ACM International Conference on Measurement and Modeling of Computer Systems, SIGMETRICS* (1996), pp. 68–79
84. R. Wunderlich, T. Wenisch, B. Falsafi, J. Hoe, SMARTS: accelerating microarchitecture simulation via rigorous statistical sampling, in *Proceedings of the 30th Annual International Symposium on Computer Architecture, ISCA* (2003), pp. 84–95
85. H. Zabel, W. Müller, A. Gerstlauer, Accurate RTOS modeling and analysis with SystemC, in *Hardware-dependent Software: Principles and Practice*, ed. by W. Ecker, W. Müller, R. Dömer (Springer, Dordrecht, 2009)
86. Z. Zhao, A. Gerstlauer, L.K. John, Source-level performance, energy, reliability, power and thermal (PERPT) simulation. IEEE Trans. Comput. Aided Des. Integr. Circuits Syst. (TCAD) (2016)
87. J. Zhu, D.D. Gajski, An ultra-fast instruction set simulator. IEEE Trans. Very Large Scale Integr. Syst. **10**(3), 363–373 (2002)
88. G. Zimmermann, The MIMOLA design system a computer aided digital processor design method, in *Proceedings of the 16th Design Automation Conference, DAC* (1979), pp. 53–58
89. V. Zivojnović, S. Tjiang, H. Meyr, Compiled simulation of programmable DSP architectures. J. VLSI Signal Process. Syst. **16**(1), 73–80 (1997)

Chapter 2
Model-Based Design and Automated Validation of ARINC653 Architectures Using the AADL

Jérôme Hugues and Julien Delange

Abstract Safety-Critical Systems as used in avionics systems are now extremely software-reliant. As these systems are life- or mission-critical, software must be carefully designed and certified according to stringent standards. One typical pitfall of corresponding development project is the late detection of safety issues or bugs at integration time that impose to redo development steps. Model-Based Engineering aims at capturing system concerns with specific notations and use models to drive the development process through all its phases—design, validation, implementation and ultimately, certification. Through a single consistent notation, such an approach would avoid undefined assumptions and traditional hurdles due to informal, text-based, specifications. In this chapter, we present recent contributions we pushed forward in the AADL architecture description language for the design and validation of Integrated Modular Avionics systems. First, we review modeling patterns to support abstractions for Integrated Modular Avionics systems. We then introduce capabilities to check all ARINC653 patterns are enforced at model-level. In addition, we review error modeling and safety analysis capabilities towards the production of safety reports conforming to ARP4761 recommendations, along with code generation strategies to map model elements to code. All these contributions are integrated in one uniform modeling process based on the AADL.

Keywords AADL · EMV2 · Safety analysis · Code generation · ARINC653

J. Hugues
Institut Supérieur de l'Aéronautique et de l'Espace,
Universit de Toulouse, 31055 Toulouse, France
e-mail: jerome.hugues@isae.fr

J. Delange (✉)
Carnegie Mellon Software Engineering Institute, Pittsburgh, USA
e-mail: jdelange@sei.cmu.edu

© Springer Nature Singapore Pte Ltd. 2017
S. Nakajima et al. (eds.), *Cyber-Physical System Design from an Architecture Analysis Viewpoint*, DOI 10.1007/978-981-10-4436-6_2

2.1 Introduction

Safety-Critical Systems (as the ones used in avionics, aerospace or automotive domains) are becoming extremely software-reliant. Boeing's new 787 Dreamliner contains more than 6.5 millions lines of software code [6]. Between 2006 and 2012, the software of the F-35 increased from 6800 K to 24000 KSLOCS [16].

While this trend brings many benefits such as ease of update or upgrade, re-use of software across different product lines; it also introduces new challenges.

As software is being updated and upgraded, it becomes more complex with many collocated functions on several processors that may interact (i.e. bus connections, interference between collocated tasks). In such environment, a single software error might have significant impacts: a report claims that 50% of car warranty costs are now related to electronics and embedded software [6].

In the avionics domain, a software error can have dramatic consequences. Such systems must be carefully engineered according to stringent standards such as DO178C [19], which mandates analysis, testing and certification activities. Unfortunately, ensuring compliance with this standard is labor-intensive and costly. As the development process is mostly manual and paper-driven, many errors are introduced. This adds significant rework efforts, cost and postpones product delivery.

During the last two decades, new standards have been defined to facilitate the development of safety-critical systems. In the avionics community, the ARINC653 standard [1] focuses on isolating software in partitions so that they meet higher safety requirements while reducing the number of CPUs. Such approaches help system designers structuring their architectures; but they still need to validate software isolation and deliver assurance of implementation correctness. As development activities are loosely coupled, many errors are introduced early on and impact other activities.

As a result, efforts made in early development phases, such as requirements and architecture, spread over the development process so that more than 60% of development efforts are focused on implementation and testing [7].

This motivates a Model-Based development approach for avionics software. Models use a formal representation of the system and constitute the core of the development workflow: they are processed across the development process to validate, implement and test the system. In order to implement such a development process, designers need an appropriate language to represent system architecture with their specific characteristics using an appropriate modeling language. Using a formal language reduces errors from textual specifications and undefined assumptions while automating the development process ensures that the system is correctly validated and implemented according to the specifications.

This chapter presents our recent contributions to the Architecture Analysis and Design Language (AADL) [21] for modeling and validating avionics architectures. Through a motivational example, we illustrate how model-based help detecting safety errors early while providing support for assurance cases.

We first illustrate how to represent Integrated Modular Avionics (IMA) architectures using the ARINC653 annex [22] so as to capture specific requirements (such as isolation properties). In addition, we present our validation rules that analyze models and ensure correctness of requirements enforcement. We then focus on safety analysis, and the production of various reports mandated by safety assessment authorities, as well as code generation strategies targeting avionics-grade operating systems.

These combined validations and automated report and code generation facilities are the foundations to validate the architecture and ultimately generate safety reports, paving the path towards system airworthiness at model-level.

2.2 Boeing 777 ADIRU Case Study

On 1 August 2005 a serious incident involving Malaysia Airlines Flight 124, occurred when a Boeing 777-2H6ER flying from Perth to Kuala Lumpur also involved an ADIRU fault resulting in uncommanded maneuvers by the aircraft acting on false indications [2]. The ATSB (Australian Safety Authority) found that the main probable cause of this incident was a latent software error that allowed the ADIRU to use data from a failed accelerometer. The ATSB report [18] indicates that a model of Software Health Management (SHM) for the Boeing 777 Air Data Inertial Reference Unit (ADIRU) is involved in the incident; and provides a full explanation of the incident.

The Architecture of the Boeing 777 ADIRU (Fig. 2.1) has multiple levels of redundancy. Two ADIRU units are used, primary and secondary. The primary ADIRU consists of 4 Fault Containment Areas (FCA). Each FCA contains multiple Fault Containment Modules (FCM). The ADIRU system can continue to work without maintenance if only one fault appears in each FCA. The system can still fly with 2 faults, but it needs maintenance before next flight. The Secondary Attitude Air Data Reference Unit (SAARU) also provides inertial data to flight computers. The flight computers use the middle value between the data provided by the ADIRU and SAARU.

In the report, it was revealed that accelerometer number 5 had failed in June 2001 and could still produce high acceleration values or voltages that were erroneous. This failure was identified and accelerometer number 5 was excluded from use in acceleration computation by ADIRU subsequently. On the day of the accident ADIRU went through a power cycle. Afterwards a second accelerometer failed (number 6). This failure was identified and accelerometer number 6 was excluded. But unexpectedly the software allowed the previous failed accelerometer number 5 to be used in acceleration computation, so the high value acceleration data was produced and output to the flight computer. Then the accident occurred. The reason why the failed accelerometer number 5 was reused in acceleration computation is the ADIRU software error in the algorithm that failed to recognize accelerometer number 5 as unserviceable after a power cycle.

Such an error results from incomplete system-level engineering that did not foresee such—apparently—basic situation. We claim these can be avoided by using

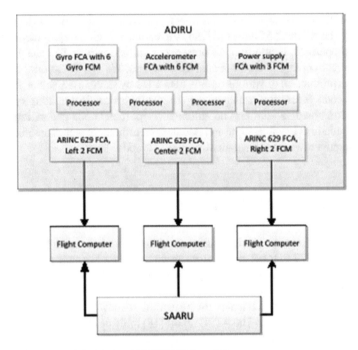

Fig. 2.1 Architecture of the Boeing-777 ADIRU, from [2]

rigorous, tool-supported engineering process. In the following, we revisit this example, introduce the core AADL language and detail the specific modeling patterns from the ARINC653 annex [22] to represent avionics architectures. We then detail how we use and extend the Resolute [15] language to validate IMA requirements in AADL models and generate assurance case.

2.3 AADL and Patterns for IMA System

In this section, we review the AADL core language and extensions we proposed as part of the SAE AS2-C committee to model avionics systems based on the Integrated Modular Avionics paradigm.

2.3.1 The AADL Core Language

The Architecture Analysis and Design Language (AADL) [21] is a modeling language standardized by SAE International. It defines a notation to model software components, the deployment/configuration on a hardware platform, and its interaction with a physical system within a single and consistent architecture model. The core language specifies several categories of components with well-defined semantics. For each component, one defines both a component type to represent its external interface; along with one or more component implementations.

For example, the task and communication architecture of the embedded software is modeled with threads and processes that are interconnected with port connections, shared data access and remote service calls.

The hardware platform is modeled as an interconnected set of buses and memory components, with virtual processors representing hierarchical schedulers, and virtual buses representing virtual channels and protocol layers. A device component represents a physical subsystem with both logical and physical interfaces to the embedded software system and its hardware platform.

System components are used to organize the architecture into a multi-hierarchy. Users model the dynamics of the architecture in terms of operational modes and different runtime configurations through the mode concept.

Users can further characterize standardized component properties, e.g., by specifying the period, worst-case execution time for threads. The language is extensible; users may adapt it to their needs using two mechanisms:

1. **User-defined properties** New properties can be defined by users to extend the characteristics of the component. This is a convenient way to add specific parameters in the model (for example, criticality of a subprogram or task)
2. **Annex languages** Specialized languages can be attached to AADL components to augment the component description through additional characteristics or requirements (for example, specifying the component behavior [14] by attaching a state-machine). They are referred to as annex languages, meaning that they are extensions to a component.

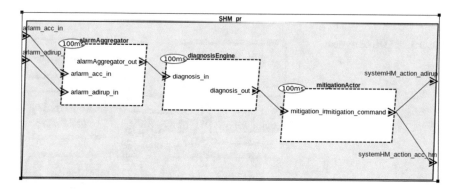

Fig. 2.2 ADIRU AADL model example - graphical representation

AADL provides two views to represent models:

1. **The graphical view** outlines components hierarchy and dependencies (bindings, connection, bus access, etc.). While it does not provide all details, this view is useful for communication and documentation purposes.
2. **The textual view** shows the description, with component interfaces, properties and annexes. It is appropriate for users to capture internal details and for tools to process and analyze the system architecture from models.

Figure 2.2 represents a simple AADL model (excerpt from the ADIRU case study) that contains three threads that communicate together. The corresponding textual notation of this model is shown in Listing 2.1.

```
process implementation systemHM_process.impl
subcomponents
    alarmAggregator: thread threads::alarmAggregator.impl;
    diagnosisEngine: thread threads::diagnosisEngine.impl;
    mitigationActor: thread threads::mitigationActor_th.impl;
connections
    C1: port arlarm_acc_in<->alarmAggregator.arlarm_acc_in;
    C2: port arlarm_adirup_in<->alarmAggregator.arlarm_adirup_in;
    -- [..]
end systemHM_process.impl;
```

Listing 2.1 ADIRU AADL example model - textual representation

The AADL model annotated with properties and annex language clauses is the basis for analysis of functional and non-functional properties along multiple dimensions from the same source, and for generating implementations, as shown in Fig. 2.3. We discuss existing analysis tools and methods to analyze and validate AADL models in Sect. 2.4.

Fig. 2.3 AADL ecosystem

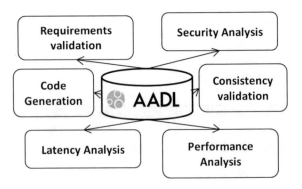

2.3.2   Modeling Integrated Modular Architectures with AADL

The goal of the Integrated Modular Avionics (IMA) concept is to integrate software components in common hardware modules. The IMA approach aims at deploying the same portable software component on different execution hardware and to change their configuration or communication policy without impacting functional aspects (and the associated code). However, such design flexibility requires capability to analyze and verify the architecture to ensure that application requirements (such as deadline, memory or safety) are met. For example, relocating a software component from one processor to another affects the scheduling and latency. Having the ability to capture this change, analyze the architecture and validate these requirements will simplify the design of such architectures.

The AADL language can be leveraged to capture IMA architectures, its software aspects and its configuration and deployment policy. For description of software aspects, AADL provides the following components:

- **data** components represent types used on components interfaces (AADL `data` and `event data port`) to communicate values between functions.
- **subprogram** components capture logical software units executed by an ARINC653 process (AADL `thread` component). Such components can be realized using traditional languages (e.g. Ada or C) or functional models (e.g. Simulink or SCADE).
- **thread** components specify a task executing code (AADL `subprogram`). This concept is similar to a POSIX thread (also called process by some standard such as ARINC653). Timing requirements of AADL `thread` components (e.g. period, deadline, execution time) are specified using the core properties.
- **process** components provide a separated memory space hosting several tasks (AADL `thread` components); UNIX processes.

For configuration and deployment, AADL includes the following components:

- **bus** components represent the physical (e.g. wires) and logical (e.g. protocol) connections (i.e. ethernet bus) that transport data across physical nodes.
- **memory** components capture a physical memory (i.e. RAM) and its logical decomposition (i.e. memory segments).
- **processor** components represent a physical processor as well as the execution runtime (i.e. operating system such as Linux or, in the context of IMA systems, an ARINC653 module).

These components are integrated to capture the software platform and its execution runtime. The deployment policy can be easily modified in the model by changing component associations so that engineers can evaluate pros and cons of different configurations. Ultimately, AADL models are analyzed by tools to evaluate the architecture metrics (e.g. latency [11] or scheduling [8]). Let us note this approach is currently transitioning to the industry [17].

The core AADL language provides an accurate semantics to capture IMA principles (deployment of software components on different execution architectures, components reuse, etc.). However, as such, it does not support the representation of some specific characteristics of IMA architecture, in particular, the Operating System. IMA systems use an ARINC653-compliant Operating System, a time and space partitioning system that separates software into partitions. While AADL provides the capability to represent such Operating System, it still needs to be extended with specific properties and modeling patterns. Next sections introduce the AADL ARINC653 annex, a standardized document that specifies how to capture ARINC653-compliant systems with the AADL.

2.3.3   The AADL ARINC653 Annex

This section introduces the AADL ARINC653 annex [22] that provides the ability to model ARINC653-compliant architectures. Models that are compliant with this annex are conformant with the IMA principles but also specific ARINC653 constraints, such as time or space isolation.

The AADL ARINC653 annex [22] defines modeling patterns and specific properties to represent ARINC653 [1] platform requirements. ARINC653 defines specific concepts (e.g. time and space partitioning) that require additional AADL properties to be captured. Standardizing modeling patterns provides guidance to represent ARINC653 systems so that AADL users and tools use the same modeling patterns, making model analysis portable.

The next paragraphs introduce the modeling patterns and specific properties of the AADL ARINC63 annex and discuss the recent additions made to this standard document, especially regarding the specification of the Health Monitoring policy.

2.3.3.1   ARINC653 Module

An ARINC653 module is captured using a `processor` component that represents the physical processor and the isolation layer (e.g. the isolation kernel ensuring time and space separation between partitions). This component also contains `virtual processor` components, each one representing a partition.

The `processor` defines the time isolation policy of the `ARINC653` module: time slots durations and assignments to each partition. This requirement is captured by the `ARINC653::Module_Schedule` property. Partition should have at least one time slot to be sure it is executed at each module period.

2.3.3.2   ARINC653 Partition

An ARINC653 partition is defined by two main assets:

- **application software** it is associated with the partitions as ARINC653 processes, using AADL `thread` components. It is important to distinguish the difference between an AADL `process` component and an ARINC653 process, as the same word are used in both standards with different semantics. The former represents the ARINC653 application software partition while the latter is a task and mapped to an AADL `thread` component.
- **execution runtime** resources available to execute the application software partition. It is captured using an AADL `virtual processor` associated with the ARINC653 module (AADL `processor`).

The application software (AADL `process`) is associated with an execution runtime using the AADL processor bindings (property `Actual_Processor_Binding`). It is also associated with a memory segment (i.e. space isolation) using AADL memory association bindings (property `Actual_Memory_Binding`).

2.3.3.3   Memory Configuration

In an ARINC653 architecture, the main memory is separated in several disjoint segments. Each segment is associated with one partition, ensuring space isolation between collocated partitions.

The main memory component (e.g. RAM) is captured using a `memory` component. The policy is specified decomposing this AADL component using `memory` sub-components, each one representing one segment. Memory segment characteristics (e.g. segment size, base address) are specified by attaching AADL properties the corresponding AADL `memory` components.

Then, each segment is allocated to one partition using the AADL memory binding mechanism (property `Actual_Memory_Binding`). This is specified by associating the partition application software (e.g. AADL `process` component) to the memory segment (e.g. AADL `memory` component).

2.3.3.4   Scheduling Parameters

ARINC653 mandates a hierarchical scheduling approach with two levels:

1. The **module level** schedules partitions using a fixed, predictive time-line algorithm repeated at a given rate (called the Major Frame).
2. The **partition level** schedules ARINC653 processes (AADL `thread`) within partitions. This is policy is partition-dependent and relies on the mechanisms supported by the underlying partition execution runtime.

The scheduling policy at the *module level* is specified in AADL processor components, which represent ARINC653 modules. The ARINC653::Module _Schedule property defines a list of time slots for each partition, while the partitions' scheduling rates are defined by the ARINC653::Module_Major_Frame property.

The scheduling policy at the *partition level* is defined in AADL virtual processor components which represent the partition execution runtime. The scheduling policy is captured with the AADL Scheduling_Protocol property.

2.3.3.5 Intra-partition Communications

ARINC653 Intra-Partition communication are channels between ARINC653 processes (or AADL thread components) located within the partition. Such channels are confined within the partitions and do not require any specific capability from ARINC653 module. The standard distinguishes several intra-partition mechanisms: buffers, blackboards, events and semaphores which are translated in AADL using (respectively) event data ports, data ports, event ports and shared data components.

The associated AADL data classifier associated to a port represents the type of data used by the communication channels (ARINC653 buffers and blackboards). In addition, AADL properties (e.g. Queue_Size) are attached to AADL interfaces to represent specific requirements (e.g. number of data instances within a buffer, management of history, etc.).

2.3.3.6 Inter-partitions Communications

ARINC653 inter-partitions communications specify communication channels between partitions. This type of communication must be explicitly configured by the ARINC653 operating system. Only declared channels can be created and used at runtime. Correct configuration and implementation of this mechanism ensure space isolation across partitions by avoiding data leakage between partitions classified at different assurance level.

The ARINC653 standard distinguishes two types of inter-partitions communications: queuing and sampling ports, which are mapped in AADL using (respectively) event data ports and data ports. Ultimately, partitions ports are connected to tasks in order to represent data usage and explicitly capture what software component (thread or subprogram) processes and uses it.

2.3.3.7 Health-Monitoring Policies

ARINC653 health-monitoring policy configures error detection mechanisms and associated mitigation and recovery strategy to keep the system in a safe state.

Table 2.1 Mapping rules between ARINC653 and AADL concepts

ARINC653 concept	AADL concept
Module	Processor component
Partition	Process component bound to a virtual processor component and a memory component
Space isolation	Decomposition of physical memory components into logical memory components
Time isolation	Each partition (process component) is bound to a virtual processor component that is itself bound to a processor component (representing the module)
Process	Thread component
Queuing ports	Event data ports across process components
Sampling port	Data ports across process components
Buffer	Event data port across thread components
Blackboard	Data port across thread components

The general concept is that each potential error (e.g. a divide by zero exception, inconsistent memory access, etc.) is associated with a recovery action such as restarting the partition where the fault originated. The ARINC653 standard distinguishes three levels of health monitoring: module, partition and process.

The ARINC653 AADL annex defines a simple approach to map Health Monitoring policies with two properties:

- `ARINC653::HM_Error_ID_Levels` defines the level for each system error and at which level it is detected and eventually recovered;
- `ARINC653::HM_Error_ID_Actions` defines the set of recovery actions for each error that can be detected by the ARINC653 executive.

These properties replace the approach from previous versions of the ARINC653 AADL annex which used properties at each health monitoring level. Because the health monitoring policy was spread over the component hierarchy, this former method was confusing as it might introduce non-deterministic specifications (for example, having the same error handled at several levels which is not a legal ARINC653 specification).

2.3.3.8   Mapping Rules

Between AADLv2 and ARINC653 concepts are summarized in Table 2.1.

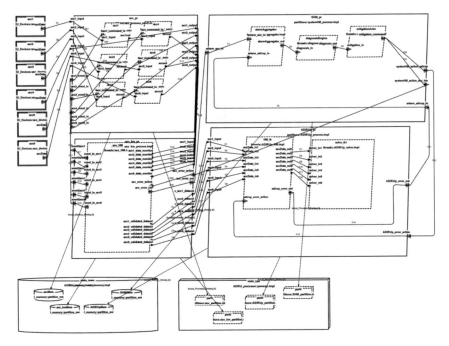

Fig. 2.4 ADIRU full model (graphical)

2.3.4 ADIRU Full Model

We captured in a set of AADL packages the full model.[1] It represents 1.5 KSLOCs of models and capture all facets of the functional architecture of the ADIRU, along with its scheduling and memory configuration parameters. A graphical representation is shown in Fig. 2.4.

Let us note this model has a high level of complexity, and captures all facets of the initial cas e study. Yet, it remains of tractable size thanks to the compact textual representation. From this model, several analyses can be performed. In the following, we focus on three of them: model-based assurance, safety analysis and code generation. For each, we explain how the model has been extended to address a particular concern.

2.4 Model-Based Assurance with AADL

The system architecture captured in AADL can be processed and validated against system requirements. Such analysis can be processed by specific tools that browse models components, extract properties and evaluate their correctness against the requirements.

---

[1] The model is available as part of the AADLib library of models: http://www.openaadl.org/aadlib. html.

This validation process is useful but is difficult to analyze, especially when there are a lot of inter-dependent results. Investigating analysis results and finding potential issues can be challenging, especially when architectures have inter-dependent requirements. In order to make the analysis review easier, we extend our analysis tool and auto-generate an assurance case from the validation results. This shows the inter-dependencies of each requirements using a hierarchical notation and details which ones are not enforced. The assurance case, associated with the validation results, constitutes an indicator of the system architecture quality (how many requirements are covered and validated in the architecture).

The next sections introduce the Resolute validation language, our extension for producing assurance case from validation results and how we apply this technique to produce assurance cases to check IMA requirements.

2.4.1   Validation of AADL Models

Analysis tools process AADL models and automatically check their correctness with regard to specific quality attributes (e.g. security, safety, performance). To date, AADL has been already successfully used to validate several quality attributes such as security [12], performance, [11] or safety [9]. Analysis functions have been designed in the Open Source AADL Tool Environment (OSATE) [5], an Eclipse-based modeling framework. Analysis tools are implemented as Eclipse plug-ins that browse the components hierarchy, retrieve informations from the components (through AADL `properties` or `annex` languages) and produce an analysis report.

However, writing analysis methods as Eclipse-plugins require learning the internals of the modeling tool, study the Eclipse platform as well as the AADL meta-model. This makes the design of new analysis features difficult for non computer-science experts, which reduce the development of new analysis capabilities. Model analysis can be implemented using general constraints language (such as OCL [20]) but such approaches are often difficult to use, complicated to use and not user-friendly [4].

```
no_double_fanin() <=
  ** " All incoming feature have only one connection" **
  forall (c : component) . true => has_single_fanin (c)


has_single_fanin (comp : component) <=
  ** "All IN features have one connection on " comp **
  forall (f : features (comp)) . (direction(f) = "in")
                    => (length (connections (f)) = 1)
```

Listing 2.2 Example of a RESOLUTE theorem - check for single fanin

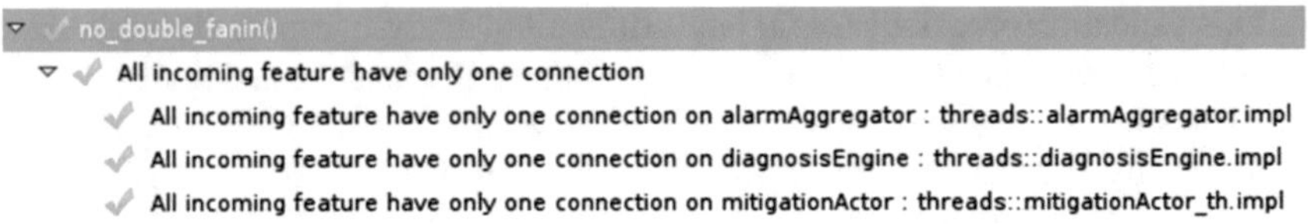

Fig. 2.5 RESOLUTE analysis result

To overcome this issue, the AADL community proposed a specific extension (through the annex mechanism of AADL), RESOLUTE [15], to process and analyze a model with a specific, user-friendly query language. It allows system designers to write new analysis methods within the modeling platform without having to learn the basics of Eclipse plug-ins development or any details of the AADL meta-model. When analyzing a model, RESOLUTE produces a hierarchical graphic representation of the execution results (as in Fig. 2.5).

Listing 2.2 shows an example of a resolute theorem that checks all incoming interfaces of all components are connected to a single source. Figure 2.5 shows the graphical representation of the analysis results when using this theorem on the AADL model introduced previously in Fig. 2.2 and Listing 2.1. In this example, the analysis passes: all incoming interface is connected to a single source The graphical representation of analysis results helps system designers to automatically check AADL models against specific requirements. A complete description of the language and analysis tools and more details can be found in [15].

2.4.2   Application to ARINC653 Requirements

We implemented ARINC653 validation rules using the RESOLUTE [15] language introduced in Sect. 2.4, and the new capability to generate assurance case [3, 13] from analysis results with a GSN notation. We designed a library of predefined validation theorems that validate ARINC653 requirements in AADL by checking that:

- Each partition AADL `process` is associated with exactly one memory segment (AADL `memory` component) and one partition execution runtime (AADL `virtual processor` component).
- Each ARINC653 module (AADL `processor` components) specify the partitions scheduling policy (property `ARINC653::Module_Schedule`) and execute each partition at least once during each scheduling period.
- All ARINC653 processes (AADL `thread`) define their scheduling characteristics (e.g. dispatch protocol, period, deadline).
- The ARINC653 Health-Monitoring Policy address all potential errors that are listed in the ARINC653 standard (such as divide by zero, application error, module error, etc.).

- Each memory segment (AADL `memory` component) is associated with at most one partition (AADL `process` component).
- All queuing ports or buffers (represented with AADL `event data ports`) specify the maximum number of data instances they can store (property `Queue_Size`).

These rules have been written in a RESOLUTE theorem library and integrated in OSATE. System designers can then use them directly without having to write any additional code. Note that these rules do not directly address safety issues related to the incident from Sect. 2.2 but detect any deployment or configuration issue that can led to such issue. This is part of a model-development process that can catch several type of errors. The next paragraph presents how safety analysis can be performed from the same model and detect safety issues as the one from Sect. 2.2.

2.5 Safety Analysis

Aerospace Recommended Practice (ARP) 4761 from Society of Automotive Engineers (SAE) defines a process for using common modeling techniques, and analysis methods such as Functional Hazard Assessment (FHA), Fault Tree Analysis (FTA) or Fault Impact Analysis (FIA).

As part of the AS5506/A1 standard document [22], AADL has been enhanced with capabilities to support capture of erroneous behavior and to analyse their impact on the global architecture through the Error Modeling Annex v2 (EMV2 thereafter). EMV2 supports architecture fault modeling at three levels:

- Error propagation between components and their environment: Modeling of fault sources in a system, their impact on other components or the operational environment through propagation.
  It allows for safety analysis in the form of hazard identification, fault impact analysis, and stochastic fault analysis.
- Component faults, failure modes, and fault handling: Modeling of fault occurrences within a component, resulting fault behavior in terms of failure modes, effects on other components, the effect of incoming propagations on the component, and the ability of the component to recover or be repaired.
  It allows for modeling of system degradation and fail-stop behavior, specification of redundancy and recovery strategies providing an abstract error behavior specification of a system without requiring the presence of subsystem specifications.
- Focus on compositional abstraction of system error behavior in terms of its subsystems. It allows for scalable compositional safety analysis.

In addition, EMV2 introduces the concept of error type to characterize faults, failures, and propagations. It includes a set of predefined error types as starting point for systematic identification of different types of error propagations providing an error propagation ontology. Users can adapt and extend this ontology to specific domains. See [10] for a more detailed presentation of EMV2 features.

```
device implementation acc_device.impl
annex EMV2
{**
```

use types ADIRU_errLibrary;
use behavior ADIRU_errLibrary::simple;

error propagations
  accData : out propagation{ValueErroneous};
flows
  f1 : error source accData{ValueErroneous} when failed;
end propagations;

Error definition

properties
  emv2::hazards =>   **Error documentation**
  ([crossreference => 'N/A';
     failure => "Accelerometer value error';
     phases => ("in flight");
     description => "Accelerometer starts to send an erroneous value";
     comment => "Can be critical if not detected by the health monitoring';
  ])
applies to accData.valueerroneous;

```
    EMV2::OccurrenceDistribution => [ ProbabilityValue => 3.4e-5 ; Distribution => Fixed;]
    applies to accData.valueerroneous;
**};
end acc_device.impl;
```

Fig. 2.6 Extension for safety concerns

Component	Error	Hazard Description	ossreferer	Functional Failure	Operational Phases	Comment
acc1	"ValueErroneous on accData"	"Accelerometer starts to send an erroneous value"	"N/A"	"Accelerometer value error"	"in flight"	"Can be critical if not detected by the health monitoring"
acc2	"ValueErroneous on accData"	"Accelerometer starts to send an erroneous value"	"N/A"	"Accelerometer value error"	"in flight"	"Can be critical if not detected by the health monitoring"
acc3	"ValueErroneous on accData"	"Accelerometer starts to send an erroneous value"	"N/A"	"Accelerometer value error"	"in flight"	"Can be critical if not detected by the health monitoring"
acc4	"ValueErroneous on accData"	"Accelerometer starts to send an erroneous value"	"N/A"	"Accelerometer value error"	"in flight"	"Can be critical if not detected by the health monitoring"
acc5	"ValueErroneous on accData"	"Accelerometer starts to send an erroneous value"	"N/A"	"Accelerometer value error"	"in flight"	"Can be critical if not detected by the health monitoring"
acc6	"ValueErroneous on accData"	"Accelerometer starts to send an erroneous value"	"N/A"	"Accelerometer value error"	"in flight"	"Can be critical if not detected by the health monitoring"

Fig. 2.7 FHA report generated

We extended the previous model to add EMV2 concepts. Thanks to AADL encapsulation and inheritance mechanism, we can separate regular interfaces and properties from safety concerns (see Fig. 2.6).

By defining specific error types, propagations across components and error occurence probabilities, we can generate directly through OSATE2 analysis plug-ins both Fault Hazard Analysis (FHA) that captures the list of hazards that are derived from each function (Fig. 2.7) and Fault Trees (Fig. 2.8, note the figure is symetrical for the 6 accelerometers, we cropped it to make it legible) that represent the combinations of errors that could lead to a system-level failure.

Through Fault Impact and Faul Tree analyses, we could replay all scenarios depicted in the original case study, demonstrating the expressive power of EMV2 to capture in a concise way error propagation.

2.6 From Model to Code

The complete ADIRU system has been modeled in AADL thanks to the materials provided by the ATSB. To this model, we applied the validation scheme presented in Sect. 2.4.2 to the model, so as to validate the model against the ARINC653 requirements, and then perform safety analysis.

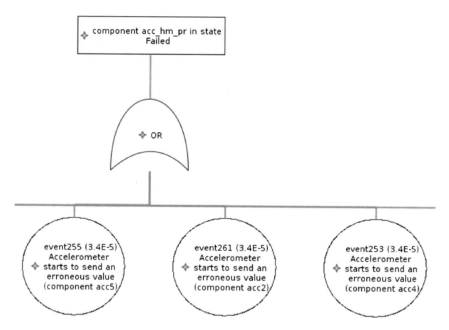

Fig. 2.8 FTA report generated

Let us note the AADL model captures all configuration parameters: partitions, buffers, link to memory segments, etc. We leveraged this information to validate the model is sound in the previous section. From this description, we can actually go further and also generate the corresponding source code.

To do so, we enhanced the Ocarina code generator [23] to target ARINC653 Application Executive (APEX), and more specifically DDC-I DeOS and WindRiver VxWorks653. These are extensions to previous work we did when targetting the FLOOS POK partitioned kernel. We rely on the Code generation annex that has been published in [22]. This annex provides guidelines to connect user-provided source code to a full distribution middleware generated from an AADL model.

Code generation covers two facets of the ARINC653 process:

1. Generation of XML descriptors: these configuration files describe all the resources required by the APEX to run: name and configuration parameters. These are derived from the definition of AADLv2 memory components that define the memory layouts, virtual processors capturing partitions and regular AADL threads, processes and subprograms;
2. Generation of code skeletons to be populated by the user source code. This codes provides regular patterns for periodic or sporadic activations, along with patterns for inter-/intra-partition communications using ARINC653 queueing and sampling ports.

Let us note that at this level, the initial AADLv2 model can be reused almost as-is. The only necessary addition is a property that indicates the APEX we want to target. This is captured in a property that is specific to our toolset.

Hence, from the AADL model, one can now generate a full prototype of the running application for both simulation and run-time validation purposes, combined with model-level validation. As such artefacts are usually produced manually by system engineers using (mostly) textual specification, generating them from a consistent semantics modeling language as AADL exhibits the following benefits:

1. **Costs reduction** as assurance cases can be automatically produced from models, it avoids labor-intensive costs to create them, especially considering the size of such documents on real systems. Also, because of this automation capability, the system's middleware code source can be updated and maintained as soon as the system is modified.
2. **Higher fidelity** using a semi-formal language (such as AADL) removes the errors related to informal specifications. It ensures the quality of the generated artifact and removes any indeterminism related to undefined assumptions made while reading non-formal, textual system specifications.

From the code generated, and its execution in a simulated environment, one can demonstrate all the scenarios that were initially tested as part of the safety analysis: simulation of the failure of a sensor, correct re-configuration of the system in the corrected revision and complete the performance analysis and evaluate the worst-case response time before complete recovery.

2.7   Conclusion and Future Work

Safety-critical software, such as the one used in avionics, must be carefully designed, validated and implemented. Such systems use dedicated execution platform (the IMA and its related ARINC653 operating system) and must comply with stringent certification standards (DO178C). Because of these requirements designing and maintaining such software is cost-intensive, and actual development methods no longer scale. Over the years, Model-Based approaches have shown interesting benefits to reduce development costs while maintaining (or even increasing) system quality.

In this chapter, we introduce our Model-Based Engineering approach to design and validate ARINC653 systems. We leverage the AADL language to capture IMA architecture and their requirements and propose specific modeling patterns for modeling ARINC653 operating systems.

In addition, we design a validation library for ARINC653 systems using RESOLUTE, an AADL-specific constraint language to analyze and validate software architectures. Using this tool, we are able to automate some system validation activities and auto-generate assurance cases, usually created manually. Such an automation might reduce manual efforts and keep certification documents up-to-date with the actual specifications.

Finally, we performed both safety analysis using AADLv2 EMV2 so that engineers can analyze fault impacts and generate safety documents (e.g. Fault-Tree Analysis, Failure Mode and Effects Analysis); and extended actual AADL code generation capabilities from the Ocarina toolset to auto-produce an implementation code that targets an ARINC653-compliant operating system.

Hence, we propose a large palette of tools to support all major activities for the engineering of safety-critical avionics system: full architecture capture, covering both functional and dysfunctional facets; analysis of system-level requirements on architecture artifacts, safety analysis and finally code generation. In our view, code generation in a simulated environment serves both for rapid prototyping and validation of many design choices.

All tools presented in this chapter are available through OSATE2 and Ocarina, and are available as free software. The ADIRU model is also fully public, to help the community better understand the close relationships between all those models. As we stated the model supports multiple verification and validation activities while remaining of modest size.

Future work will consider closer interaction with certification processes so as to align our contributions with current practices, but also challenge the benefits of model-based to reduce certification costs.

Acknowledgements Copyright 2016 Carnegie Mellon University. This material is based upon work funded and supported by the Department of Defense under Contract No. FA8721-05-C-0003 with Carnegie Mellon University for the operation of the Software Engineering Institute, a federally funded research and development center.

No warranty. This Carnegie Mellon University and Software Engineering Institute Material is furnished on an as-is basis. Carnegie Mellon University makes no warranties of any kind, either expressed or implied, as to any matter including, but not limited to, warranty of fitness for purpose or merchantability, exclusivity, or results obtained from use of the material. Carnegie Mellon University does not make any warranty of any kind with respect to freedom from patent, trademark, or copyright infringement.

[Distribution Statement A] This material has been approved for public release and unlimited distribution. Please see Copyright notice for non-US Government use and distribution.

DM-0003495.

References

1. Airlines Electronic Engineering, Avionics application software standard interface—ARINC653. Technical Report (ARINC—Aeronautical Radio, Inc., 1997)
2. ATSB Transport Safety Investigation Report, In-flight upset event 240 km north-west of Perth, WA Boeing Company 777-200, 9M-MRG. Technical Report Aviation Occurrence Report 200503722 (ATSB, 2005)
3. R. Bloomfield, P. Bishop, Safety and assurance cases: past, present and possible future an adelard perspective, in *Making Systems Safer*, ed. by C. Dale, T. Anderson (Springer, London, 2010), pp. 51–67
4. J. Cabot, R. Clarisó, UML/OCL verification in practice, in *ChaMDE 2008 Workshop Proceedings: International Workshop on Challenges in Model-Driven Software Engineering* (2008), pp. 31–35

5. Carnegie Mellon Software Engineering Institute: OSATE—Open Source AADL Tool Environment. Technical report (2016), http://www.aadl.info
6. R.N. Charette, This car runs on code, in *IEEE Spectrum*, Feb 2009
7. B. Clark, R. Madachy, *Software Cost Estimation Metrics Manual for Defense Systems* (Software Metrics Inc., Haymarket, 2015)
8. J. Craveiro, J. Rufino, F. Singhoff, Architecture, mechanisms and scheduling analysis tool for multicore time-and space-partitioned systems. ACM SIGBED Rev. **8**(3), 23–27 (2011)
9. J. Delange, P. Feiler, D. Gluch, J.J. Hudak, AADL fault modeling and analysis within an ARP4761 safety assessment. Technical Report (2014)
10. J. Delange, P.H. Feiler, Architecture fault modeling with the AADL error-model annex, in *40th EUROMICRO Conference on Software Engineering and Advanced Applications, EUROMICRO-SEAA 2014*, Verona, Italy, 27–29 Aug 2014 (2014), pp. 361–368
11. J. Delange, P.H. Feiler, Incremental latency analysis of heterogeneous cyber-physical systems, in *Proceedings of 3rd IEEE International Workshop on Real-Time and Distributed Computing in Emerging Applications, REACTION 2014*, Rome, Italy, 2 Dec 2014 (2014)
12. J. Delange, L. Pautet, F. Kordon, Design, implementation and verification of MILS systems. Softw. Pract. Exper. **42**(7), 799–816 (2012)
13. E. Denney, G. Pai, J. Pohl., Advocate: an assurance case automation toolset, in *Proceedings of the 2012 International Conference on Computer Safety, Reliability, and Security, SAFECOMP 2012* (Springer, Berlin, Heidelberg, 2012), pp. 8–21
14. R. Frana, J.-P. Bodeveix, M. Filali, J.-F. Rolland., The AADL behaviour annex – experiments and roadmap, in *Engineering Complex Computer Systems* (2007), pp. 377–382
15. A. Gacek, J. Backes, D. Cofer, K. Slind, M. Whalen, Resolute: an assurance case language for architecture models, in *Proceedings of the 2014 ACM SIGAda Annual Conference on High Integrity Language Technology* (ACM, 2014), pp. 19–28
16. C. Hagen, J. Sorensen, Delivering military software affordably, in *Defense AT&L* (2013), pp. 30–34
17. A.V. Khoroshilov, I. Koverninskiy, A. Petrenko, A. Ugnenko, Integrating AADL-based tool chain into existing industrial processes, in *ICECCS* (2011), pp. 367–371
18. N. Mahadevan, A. Dubey, G. Karsai, A case study on the application of software health management techniques. ISIS-11-101, Jan 2011 (2011)
19. Military Aerospace, DO-178C nears finish line with credit for modern tools and technologies, May 2010
20. OMG, *UML 2.0 Specification* (Object Management Group, Final Adopted Specification, 2005)
21. SAE International, *AS5506B—Architecture Analysis and Design Language (AADL)*, Sept 2012
22. SAE International, *AS55061/A—SAE Architecture Analysis and Design Language (AADL) Annex Volume 1*, Oct 2015
23. B. Zalila, I. Hamid, J. Hugues, L. Pautet, Generating distributed high integrity applications from their architectural description

Chapter 3
Formal Semantics of Behavior Specifications in the Architecture Analysis and Design Language Standard

Loïc Besnard, Thierry Gautier, Paul Le Guernic, Clément Guy, Jean-Pierre Talpin, Brian Larson and Etienne Borde

Abstract In system design, an architecture specification or model serves, among other purposes, as a repository to share knowledge about the system being designed. Such a repository enables automatic generation of analytical models for different aspects relevant to system design (timing, reliability, security, etc.). The Architecture Analysis and Design Language (AADL) is a standard proposed by SAE to express architecture specifications and share knowledge between the different stakeholders about the system being designed. To support unambiguous reasoning, formal verification, high-fidelity simulation of architecture specifications in a model-based AADL design workflow, we have defined a formal semantics for the behavior specification of the AADL, the presentation of this semantics is the aim of this chapter.

Keywords Architecture modeling · Formal semantics · Synchronous concurrency · Code generation · AADL

3.1 Introduction

In system design, an architecture specification serves several important purposes. First, it breaks down a system model into manageable components to establish clear interfaces between them. In this way, complexity becomes manageable by hiding details that are not relevant at a given level of abstraction. Clear, formally defined, component interfaces allow us to avoid integration problems at the implementation

L. Besnard
IRISA, Rennes, France
e-mail: loic.besnard@irisa.fr

T. Gautier · P. Le Guernic · C. Guy · J.-P. Talpin (✉)
INRIA, Rennes, France
e-mail: jean-pierre.talpin@irisa.fr

B. Larson
FDA Scholar, Kansas State University, Manhattan, KS, USA

E. Borde
Telecom ParisTech, Paris, France

© Springer Nature Singapore Pte Ltd. 2017
S. Nakajima et al. (eds.), *Cyber-Physical System Design from an Architecture Analysis Viewpoint*, DOI 10.1007/978-981-10-4436-6_3

phase. Connections between components, which specify how components affect each other, help propagate the effects of a change in one component to the linked components.

More importantly, an architecture model is a repository to share knowledge about the system being designed. This knowledge can be represented as requirements, design artefacts, component implementations, held together by a structural backbone. Such a repository enables automatic generation of analytical models for different aspects relevant to system design, such as timing, reliability, security, performance, energy, etc. Since all the analyses are generated from the same source, the consistency of assumptions w.r.t. guarantees, of abstractions w.r.t. refinements, used for different analyses, becomes easier, and can be properly ensured in a design methodology based on formal verification and synthesis methods.

Several standards for modeling embedded architectures have emerged in recent years: the SAE AADL[1] [1], SysML,[2] and UML MARTE [17]. Each of them represents different design approaches, embodies different concepts, and serves different purposes. We focus on the AADL, and the scope and precision of concepts defined by this standard, to define a formal semantics for a significant subset of its behavioral specification annex language, often called 'BA'. Just as non-functional properties (timing, performance, energy, security properties), such descriptions can be attached to threads, processes, or any object of the standard (bus, sensor, actuator, port) to formally specify its behavior, as specified in the standard (e.g. a bus), or refine it (e.g. as an AFDX bus).

Since it began being discussed in the AADL standard committee, the formal semantics defined in this article evolved from a synchronous model of computation and communication [4] to a semantic framework for time and concurrency in the standard: asynchronous, synchronous or timed, to serve as a reference for model checking, code generation or simulation tools uses with the standard. These semantics are simple, relying on the structure of automata already present in the standard, yet provide tagged, trace semantics framework to establish formal relations between (synchronous, asynchronous, timed) usages or interpretations of behavior.

3.2   Example of an Adaptive Cruise Control System

To illustrate the definition and use of a formal semantics for the AADL behavior annex, we consider the case study of an Adaptive Cruise Control (ACC) system, Fig. 3.1.

ACC systems implement two main functions:

1. an ACC can automatically sustain a preset speed (as a conventional Cruise Control system), and

---

[1] http://www.aadl.info/aadl/currentsite/.

[2] http://www.omg.org/spec/SysML/1.4/.

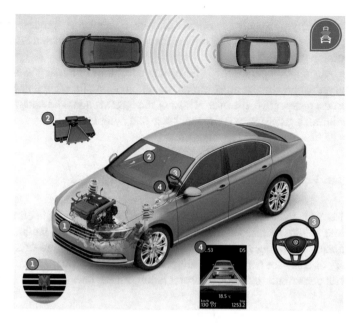

Fig. 3.1 Adaptative Cruise Control

2. an ACC can adapt the vehicle's speed to maintain a safe distance with other vehicles so as to prevent collisions.

To implement these functions, the ACC requires data from different sensors: speedometer, laser, radar (1, 2) to detect vehicles or obstacles ahead, and wheel sensor to adjust the focus point of the laser or radar. The ACC receives commands from the driver through buttons allowing to set the preferred speed and to activate or deactivate the system (3, 4).

Depending on the situation (presence of an obstacle or not, activation of the cruise control or not), the ACC computes the acceleration and deceleration for the vehicle to reach the needed speed: the preferred speed of the driver if there is no obstacle and the cruise control is on, or the speed of the vehicle ahead. Finally, it acts on the vehicle speed through its brakes and throttle.

An ACC is a safety-critical system. Hence, in addition to meeting its functional requirements, its design must satisfy design correctness objectives that concern several aspects specified in its architecture model:

- from the timing and scheduling perspective, all threads must meet their deadlines;
- reaction to the presence of an obstacle must be done in a minimum amount of time;
- from the logical perspective, the system must be free of deadlock and race condition;

- from the security perspective, critical software components (processes or systems) must be protected from less critical components, thus executed on dedicated processors;
- from the consumption perspective, the system must draw minimal power from the car battery, thus processors must run on the minimal possible frequency;
- from the cost perspective, the overall cost of the system should be minimal, which means minimizing hardware component size and complexity.

3.3 Architecture Analysis and Design Language

AADL [1] is SAE International standard AS5506C, dedicated to modeling embedded real-time system architectures. As an architecture description language, based on a component modeling approach, AADL describes the structure of systems as an assembly of software components allocated on execution platform components together with constraints and properties, including timing ones.

3.3.1 Architecture

In AADL, three distinct families of components are provided:

- software application components which include process, thread, thread group, subprogram, and data components,
- execution platform components that model the hardware part of a system including (possibly virtual) processor, memory, device, and (possibly virtual) bus components,
- composite components (systems).

Figure 3.2 presents an overview of an ACC system, consisting of:

- devices, such as sensors (speedometer, radar, wheel sensor), console with buttons and display, throttle and brakes;
- buses allowing subsystems to communicate with each other and with devices;
- controller and console subsystems.

Each subsystem in Fig. 3.2 consists of hardware components, such as processors, memories and buses; and software components: processes containing threads. Figure 3.3 presents the controller subsystem and its components: one processor, one memory, one bus connecting the processor and the memory and one controller process. The controller process itself contains four threads, one for each sensor, and the ComputeActionThread, which is responsible for sending speed up, slow down or complete stop signals to the throttle and brakes of the vehicle.

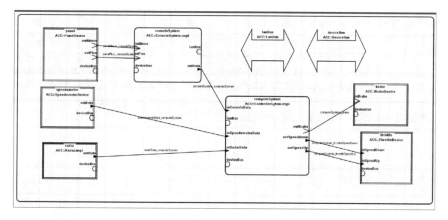

Fig. 3.2 Overview of the Adaptive Cruise Control system modeled with AADL. *Double-lined rectangles* represent devices, *double-arrows* buses and *rectangles* with *rounded corners* systems and subsystems

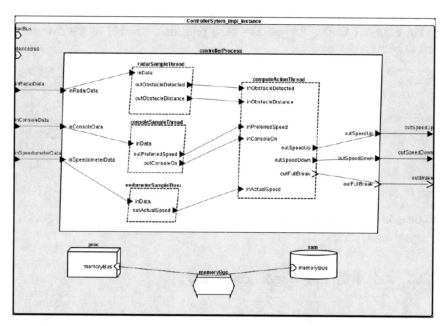

Fig. 3.3 Controller subsystem of the Adaptive Cruise Control system modeled with AADL. *Rectangles* represent processors, *double-arrows* buses, *cylinders* memories and *rhombuses* processes and threads

The AADL components communicate via data, event, and event data ports. In Fig. 3.2, ports are represented by arrows, and connections between ports by lines. Data ports are represented using filled arrowheads and event ports using empty arrowheads.

Each component has a type, which represents the functional interface of the component and externally observable attributes. Each type may be associated with zero, one or more *implementation(s)* that describe the contents of the component, as well as the *connections* between components.

3.3.2 Properties

AADL properties provide various information about model elements of an AADL specification. For example, a property `Dispatch_Protocol` is used to provide the dispatch type of a thread. Property associations in component declarations assign a particular property value, e.g., `Periodic`, to a particular property, e.g., `Dispatch_Protocol`, for a particular component.

For example, Listing 3.1 presents such properties attached to the `Compute ActionThread` thread.

```
thread implementation ComputeActionThread.impl
  properties
    -- periodic thread
    Dispatch_Protocol => Periodic;
    Period => 50 ms;
    -- thread deadline
    Deadline => 40 ms;
    -- thread WCET
    Compute_Execution_Time => 20 ms;
end ComputeActionThread;
```

Listing 3.1 Timing and scheduling properties of the ComputeActionThread thread implementation

3.3.3 AADL Timing Execution Model

Threads are dispatched periodically, triggered by the arrival of data or events on ports, or from the arrival of a subprogram call (from another thread), depending on the thread type. Three event ports are predeclared: `dispatch`, `complete` and `error` (Fig. 3.4).

A thread is activated to perform a computation at *start* time, and has to be finished before the *deadline*. A *complete* event is sent at the end of the execution. The received inputs are frozen (copied for reading) at a specified time (*Input_Time*), by default the *dispatch* time. This implies that the content of a dispatched port does not change during the execution of a thread dispatch, even though the sender may send new values in its input FIFO queue. For example, values 2 and 3 (Fig. 3.4) arriving after the first

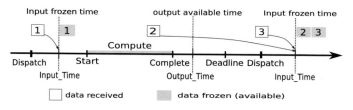

Fig. 3.4 Execution time model for an AADL thread

Input_Time will not be processed until the next *Input_Time*. As a result, the performed computation is not affected by a new input arrival until an explicit request for input (another dispatch). Similarly, the output is made available to other components at a specified point of *Output_Time*, by default at *complete* (resp., *deadline*) time if the associated port connection is immediate (resp., delayed) communication.

3.4 A Formalization Using Constrained Automata

We define the model of computation and communication of a behavior specification by the synchronous, timed or asynchronous traces of automata with variables [18]. These constrained automata are derived from *polychronous automata* defined within the polychronous model of computation and communication [12]. Automata define a behavior using transitions. A transition is composed of an initial state, a guard, an action, a final state. The guard and action of a transition are defined using logical formulas. The logical formula of the guard must be true for the transition to occur.

3.4.1 Vocabulary

Logical formulas are defined on the vocabulary W of the states S, variables V, connections and ports P defined in the lexical scope of the denoted AADL object, and of AADL constants. An identifier w in W has a type $T = typeof(w)$ and is valuated in the corresponding domain $\mathbb{D}_T$, e.g., Booleans, integers or reals, $\mathbb{D} \supseteq \mathbb{B} \cup \mathbb{Z} \cup \mathbb{R}$.[3]

We write $\mathbb{D}_x$ for the value domain of a typed identifier x. The domain of a port identifier p of type T is defined by $\mathbb{D}_p = \mathbb{D}_T^{\perp} = \mathbb{D}_T \cup \{\perp\}$. The bottom sign $\perp$ denotes the absence of a value at the given step of execution. A port value is said absent if the port is not frozen and its value is neither read or written.

---

[3]Although BA supports other types (strings, enumerations, records, arrays) our formalization focuses on numbers and Booleans without loss of generality.

3.4.2 Formulas

The set of typed formulas F_W on the vocabulary W is an algebraic set of terms that denotes the conditions, actions and constraints of an AADL object of vocabulary W. It is defined by induction from:

- Constants 0 (*false*), to mean "never", and 1 (*true*), to mean "always" (up to W).
- Atoms w of W, to mean the value of an identifier w.
- Unitary expressions:

  - $\hat{\ }p$ is the clock of p: a Boolean that denotes the presence of a value on a frozen port p, i.e., $p \neq \bot$;
  - $@p$ is the date of p: a real number that denotes the time of an event present on a port p;
  - v' denotes the next value of a variable v;
  - $\neg f$ denotes the complement of formula f, for all f in F_W.

- Binary expressions $f\ op\ g$:

  - for all Boolean formula f, g in F_W and Boolean operators $\vee$, $\wedge$, $\Rightarrow$, etc. (in particular, $f - g = f \wedge \neg g$);
  - for all numerical formula f, g in F_W and numerical operators $+, -, *, /, \%, =, <$, etc.

A formula f is the denotation of a well-typed AADL condition or action. It is hence assumed to be a well-typed, multi-sorted, logical expression. Ill-typed expressions do not define formula.

Example The formula $\hat{\ }a \wedge \hat{\ }b = 0$ stipulates that the ports a and b should never carry a value (sent or dispatched) at the same logical period of time. In the AADL, it refers to the condition "*on dispatch a*" of an object possibly triggered by a or b and allows to make the status of a as being explicitly frozen and b not (or, alternatively, empty, with "*not b'fresh*" or "*not b'count = 0*").

Conversely, $\hat{\ }a = \hat{\ }b$ expresses the synchrony of a and b at any step of execution. It can be refined by the real-time constraint $d \leq @a$, $@b < d + p$, where d is the date of the behavior's dispatch and p its period.

3.4.3 Model

A model m is a function $W \rightarrow \mathbb{D}_W$ from a vocabulary W to its domain of valuation $\mathbb{D}_W$ that is true for a formula f of F_W, written $m \models f$. A timed model $m^@$ is a function $W \rightarrow \mathbb{R} \times \mathbb{D}_W$ associating also each event with a date, that the formula must satisfy as well.

3.4.4 Automaton

The meaning of a behavior annex is defined by an incomplete automaton with variables $A = (S_A, s_0, V_A, P_A, T_A, C_A)$ where:

- S_A is the set of states of the automaton A, s_0 is the initial state.
- V_A is the set of local variables of the automaton A (V_A' designates the set of next values v' for all $v \in V_A$).
- P_A is the set of ports of automaton A, both inputs I_A and outputs O_A, $P_A = I_A \cup O_A$.
- F_A is the set of Boolean formulas of A, defined on the vocabulary $W_A = S_A \cup V_A \cup P_A$.
- The transition function $T_A \in S_A \times F_A \rightarrow F_A \times S_A$ defines the transition system of A.

  - The source formula of a transition is its guard g, defined on V_A and I_A.
  - The target formula of a transition is its action f, defined from V_A and P_A.

- $C_A \in F_A$ is the constraint of A. It must always equal 0, to mean never. It is a formula that denotes the invariants (properties, requirements) of the denoted AADL object in a form of a logical formula.

Since the variables V_A are private to the automaton A, a transition function T_A is equivalent to one in $X_A = Q_A \times F_P \rightarrow F_P \times Q_A$ over extended states $Q_A = S_A \times \mathbb{D}^{V_A}$ for all valuations $\mathbb{D}^{V_A} = \prod_{v \in V_A} \mathbb{D}_v$ of the variables V_A: for all transition $(s, g, f, d) \in T_A$, for all model $m \in (V_A + V_A') \rightarrow \mathbb{D}^{V_A}$, we have $((s, m(V_A)), m(g), m(f), (d, m(V_A'))) \in X_A$ and, for all $v' \in V_A'$ undefined in f, $m(v') = m(v)$.

Example A behavior alternating between two states of receiving a's and b's can be represented (Fig. 3.5) by a transition system defined on the vocabulary $\{a, b\}$ with two complete states s_0 and s_1 (complete states are observable states—see Sect. 3.5) and two transitions:

- $(s_0, \char"02C6 a - \char"02C6 b, true, s_1)$ denotes the transition from $s0$ to $s1$ if port a carries a value and b does not;
- $(s_1, \char"02C6 b - \char"02C6 a, true, s_0)$ denotes the transition from $s1$ to $s0$ if port b carries a value and a does not.

The role of a constraint formula such as $\char"02C6 a \wedge \char"02C6 b = 0$ is to guarantee the property ("never a and b") by all implementations of the (incomplete) automaton. For instance:

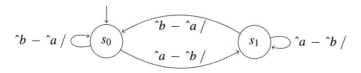

Fig. 3.5 Alternating behavior

- if an a is received in state s_1, or a b in state s_0, both the automaton and the constraint allow it: the event is consumed and the automaton remains in the same state;
- if both a and b are received in either s_0 or s_1 then the transition is denied by the constraint.

3.4.5 Properties

- The control clock 1_A of an automaton A is defined by the sum (union) of its port clocks $1_A = \sum_{p \in P_A} \hat{\ } p$.
- The trigger $tick_A(s) = \sum_{(s,g,f,d) \in T_A} (g)$ of a state s is defined by the upper bound of guard formulas g in transitions that depart from s.
- The stuttering clock of a state s is defined by $\tau_A(s) = 1_A - ((s * C_A) + tick_A(s))$. It means that an automaton A is silent in state s if and only if its model m satisfies the constraint C_A in state s and no guard can be triggered from s with m.

3.4.6 Product

The synchronous product of two automata $A = (S_A, s_0, V_A, P_A, T_A, C_A)$ and $B = (S_B, t_0, V_B, P_B, T_B, C_B)$ is defined by $A \mid B = (S_{AB}, (s_0, t_0), V_{AB}, P_{AB}, T_{AB}, C_{AB})$ with

$$S_{AB} = S_A \times S_B$$
$$V_{AB} = V_A \cup V_B$$
$$P_{AB} = P_A \cup P_B$$
$$C_{AB} = C_A \vee C_B$$
$$T_{AB} = \{((s_1, t_1), g_1 \wedge g_2, f_1 \wedge f_2, (s_2, t_2)) \mid (s_1, g_1, f_1, t_1) \in T_A \wedge (s_2, g_2, f_2, t_2) \in T_B\}$$

Product is commutative, associative, has a neutral element $(\{s\}, s, \emptyset, \emptyset, \emptyset, 0)$ and is idempotent for deterministic automata.

Example The synchronous composition of two automata A and B communicating through an immediate connection of port p can be represented by the synchronous product of A and B with the automaton representing a point-to-point one-place first-in-first-out buffer (Fig. 3.6). A queue of size n can be defined by the product of n copies of $FIFO_1$.

Fig. 3.6 $FIFO_1$

$$FIFO_1 = (\{s_0, s_1\}, s_0, \{v\}, \{p_A, p_B\}, T_{FIFO_1}, 0)$$
$$T_{FIFO_1} = \{(s_0, \hat{}p_A, v' = p_A, s_1), (s_1, true, p_B = v, s_0)\}$$

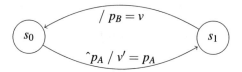

3.4.7 Small Step

The model m of a transition in an automaton A consists of a pre-condition $\text{pre}(m)$ defined on input ports $I \rightarrow \mathbb{D}_I^{\perp}$ and state variables $V \rightarrow \mathbb{D}_V$ and a post condition $\text{post}(m)$ defined on output ports $O \rightarrow \mathbb{D}_O^{\perp}$ and next values of variables $V' \rightarrow \mathbb{D}_V$.

A *small step* of an automaton A from state s to state t is defined by a model m of A that satisfies its constraint C_A, written $m \models \neg C_A$, and both the guard g and action f of a transition (s, g, f, t) of A, written $m \models g \wedge f$.

Example A small step of an automaton denotes an atomic and untimed execution step of the denoted behavior. For instance, the model $m = \{(v, 0), (v', 0), (p_A, 0), (p_B, 0)\}$ is a small step of the automaton $FIFO_1$ from s_0 to s_1: it satisfies both guard $m \models \hat{}p_A$ and action $m \models v' = p_A$.

3.4.8 Big Step

Let $n > 1$, $q_1 = (s_1, r_1)$ and $q_n = (s_n, r_n)$ two extended states of an automaton A with complete states $s_1, s_n \in S_A$ and variable valuations $r_1, r_n \in \mathbb{D}^{V_A} \simeq V_A \rightarrow \mathbb{D}_{V_A}$ (note that it may be the case that $q_n = q_1$). A *big step* of automaton A from s_1 to s_n is defined by a model $m \in P_A \rightarrow \mathbb{D}_{P_A}^{\perp}$ that, for all $1 \le i < n$ satisfies:

- $\text{pre } r_{i+1}(v) = \text{post } r_i(v')$ for all $v \in V_A$ (the *next* variable values v' at step i, $\text{post } r_i(v')$, are the regular variable values v at step $i + 1$, $\text{pre } r_{i+1}(v)$)
- $m_i = r_i \uplus m$
- $m_i \models \neg C_A$
- $(s_i, g_i, f_i, s_{i+1}) \in T_A$ and $m_i \models g_i \wedge f_i$
- $\text{pre } r_i(v) = \text{post } r_i(v')$ for all $v' \in V_A'$ not occurring in f_i and g_i
- s_i is an *execution state* if $1 < i < n$ (an *execution state* is a non observable, internal state—see Sect. 3.5).

We write $m, s_1 \models A, s_n$ to mean that m is the model of a big step of A from s_1 to s_n.

Example For instance, the model $m = \{(p_A, 0), (p_B, 0)\}$ is a big step of the automaton $FIFO_1$ from s_0 back to s_0. It abstracts the meaning of A over its port interface for the corresponding valuation of its local variables $\{(v, 0), (v', 0)\}$ that satisfies the guard and action.

3.4.9 Synchronous and Asynchronous Trace

A *synchronous trace* $B \in P_A \rightarrow (\mathbb{D}_{P_A}^{\perp})^*$ of an automaton A is a finite sequence of valuation over P_A obtained by concatenating the codomains of successive big steps. The length of B is denoted $|B|$. The set of synchronous traces of an automaton A from its initial state s_0 is defined as:

$$T(A, s_0) = \{B \in P_A \rightarrow (\mathbb{D}_{P_A}^{\perp})^* \mid 0 \leq i < |B|, \, m_i, s_i \models A, s_{i+1} \wedge \forall x \in dom(B), (B(x))_i = m_i(x)\}$$

An *asynchronous trace* $B^{\#} \in P_A \rightarrow (\mathbb{D}_{P_A})^*$ is the abstraction of a synchronous trace $B \in P_A \rightarrow (\mathbb{D}_{P_A}^{\perp})^*$ obtained by the removal of all absence marks $\perp$. For a sequence s in $(\mathbb{D}^{\perp})^*$, we denote by $s_{/\perp}$ the projection of s on $\mathbb{D}^*$. The set of asynchronous traces of an automaton A from its initial state s_0 is defined as:

$$T^{\#}(A, s_0) = \{B \in P_A \rightarrow (\mathbb{D}_{P_A})^* \mid C \in T(A, s_0) \wedge \forall x \in dom(B), B(x) = C(x)_{/\perp}\}$$

3.4.10 Timed Step and Timed Trace

A *timed step* of an automaton A from state s to state t is defined by a timed model $m^@$ defined on W_A that satisfies its constraint and the guard g and action f of a transition (s, g, f, t) of A. For all w in W_A, $m^@(w)$ refers to the value of w in $m^@$ and $m^@(@w)$ refers to the date of w in m.

A *timed trace* $B^@ \in P_A \rightarrow (\mathbb{R} \times \mathbb{D}_{P_A}^{\perp})^*$ of an automaton A is defined by the concatenation of the codomains of successive timed steps $(m_i^@)_{i \geq 0}$ of A such that for all $0 \leq i < j$, for all x in $dom(m_i^@)$, for all y in $dom(m_j^@)$, $m_i^@(@x) < m_j^@(@y)$. A timed trace $B^@$ is therefore the refinement of a synchronous trace $B \in P_A \rightarrow (\mathbb{D}_{P_A}^{\perp})^*$ associating each event in B with a date.

3.5 Behavior Annex Model

BA provides an extension to AADL to associate functional behavior specifications with AADL components. A behavior is expressed by transition systems with conditions and actions [2]. Actions can be abstract, e.g., denote the consumption of time, resources, or describe an error scenario. They can be refined to simulate and define the functional behavior of the AADL component using an imperative action language.

This section first presents how we formally express the meaning of a behavior annex (through an automaton). Then, the formal semantics of all elements of the behavior annex are defined (transition system, action and expression language, interaction protocols, etc.).

3.5.1 Formalization

Formally, the meaning of a behavior annex is defined by the axiomatic, denotational and operational interpretation of constrained, incomplete, automata with variables $A = (S_A, s_0, V_A, P_A, T_A, C_A)$ such as the one defined in Sect. 3.4. The sets S_A, V_A, P_A represent the states (including the `initial state` s_0), variables and ports of A. The guard, action and constraints of its transitions T_A and constraints C_A are denoted by multi-sorted logical formula F_A.

F_A is defined over the vocabulary W_A available in the scope of a behavior annex: AADL value constants, port, state, and variable names. They are combined using AADL logical operators and numeric operators. Operators that are specific to the model of computation and communication of a given behavior annex are $\hat{}p$, a Boolean value to mean the presence of a value on port p under synchronous interpretation (i.e., $p \neq \bot$); and $@p$, a numeric value to mean the time of an event on p, under timed interpretation.

The transition system of an automaton A is defined by the function $T_A \in S_A \times F_A \to F_A \times S_A$ whose quadruples (s, g, a, t) consist of the source state s, guard formula g, action formula a and target state t of a specified transition.

In the reminder of the section, we present each element of the behavior annex, with examples using our motivating case study, and the semantics of the element with respect to our framework.

3.5.2 Transition System

The AADL behavior annex defines a transition system (an extended automaton) described by three sections: variables declarations, states declarations, and transitions declarations. This transition system of the behavior annex is not to be confused with the transition system of the automaton interpreted to give its meaning to a behavior annex. On the one hand, we have a transition system which is part of the behavior annex, and on the other hand an automaton which is used to express the meaning of the whole behavior annex.

The automaton A of a `behavior_annex` instance is defined on the vocabulary consisting of its private variables `behavior_variable`, of its states `behavior_state`, and ports of its parent component. Its transition system T_A is the union of the transitions specified by a `behavior_transition`.

```
behavior_annex ::=
    [ variables { behavior_variable }+ ]
    [ states { behavior_state }+ ]
    [ transitions { behavior_transition }+ ]
```

We first describe how a thread can be described in the Adaptive Cruise Control model using the behavior annex. Then we present the three different sections of the transition system of the behavior annex, detailing the example. Finally, we give the formal semantics of the different elements of the transition system.

3.5.2.1   Transition System of a Thread of the Adaptive Cruise Control

In the Adaptive Cruise Control (ACC) system, the `ComputeActionThread` thread is responsible for processing the correct behavior the system should adopt (slow down, speed up or keep the speed constant) depending on the situation. Figure 3.7 pictures the transition system describing the behavior of the `Compute ActionThread` thread.

For readability sake, conditions and actions have been omitted. In the case of this transition system, conditions are tests on input signals (are they present or not) and on variables (value comparison), and actions are of two kinds: either the sending of a signal through one of the output ports of the thread; or the computation of an intermediate value, such as the vehicle speed relative to the obstacle one, or the acceleration/deceleration needed to reach a given speed, and its assignment to a variable.

The state transition system starts in the `Waiting` state, waiting for its thread to be periodically dispatched, and to pass in `Started` state. The `Waiting` state is a `complete` one, that is, a state in which a thread pauses its execution, waiting for a new dispatch.

After entering the `Started` state, depending on the inputs, the state transition system can pass in `Detected` (the system detected an obstacle) or `Console` state (the system did not detect an obstacle and the cruise control is on), or go back to the `Waiting` state (the system did not detect an obstacle and the cruise control is off).

In the `Detected` state, the system must check the emergency of the situation: if the obstacle is in an unsafe range, the system goes into the `Emergency` state and its next transition will send a signal to brakes in order to stop the vehicle; if the obstacle is outside this range, the system enters the `NoEmergency` state and then determines whether it should slow down to adapt its speed to the obstacle speed, speed up or keep the speed constant (each transition sending the corresponding signal to the throttle after the computation of the needed acceleration/deceleration). The same happens in the `Console` state depending on the current speed of the vehicle and the speed preset by the driver.

After saving useful values (e.g., current speed, current obstacle speed and distance) in the `SaveValues` state, the state transition system returns in the `Waiting` state, waiting for the next dispatch of its thread.

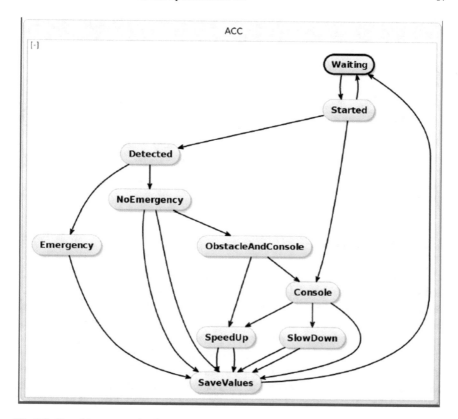

Fig. 3.7 Transition system for the `ComputeActionThread` thread

3.5.2.2 Variables Section

The variables section of the transition system of a behavior annex declares iden-
tifiers that represent variables within the scope of the behavior annex subclause.
Local variables can be used to keep track of intermediate results within the scope
of the annex subclause. They may hold the values of out parameters on subprogram calls to be made available as parameter values to other calls, as output through
enclosing out parameters and ports, or as value to be written to a data component
in the AADL specification. They can also be used to hold input from incoming port
queues or values read from data components in the AADL specification. They are
not persistent across the different invocations of the same behavior annex subclause.
Listing 3.2 presents an example of the variables section of the behavior annex of
thread `ComputeActionThread`.

```
variables
  ...
    --vv : vehicle speed (from accelerometer)
    actual_speed: Base_Types::Float;
    --vv' : previous vehicle speed
    previous_actual_speed: Base_Types::Float;
    --vo : obstacle speed (vv-vv'+(d-d'/T))
    obstacle_speed: Base_Types::Float;
    --vo' : previous obstacle speed
    previous_obstacle_speed: Base_Types::Float;
  ...
```

Listing 3.2 Sample of the variables section of the behavior annex of the ComputeActionThread thread

3.5.2.3 States Section

The states section declares all the states of the automaton. Some states may be quali-fied as initial state (thread halted), final state (thread stopped), or complete state (thread awaiting for dispatch), or combinations thereof. A state without qual-ification is referred to as execution state. A behavior automaton starts from an initial state and terminates in a final state. A complete state acts as a suspend/resume state out of which threads and devices are dispatched. Complete states thus corre-spond (with initial and final states) to the observable states of the behavior, in which computations are "paused", inputs read and outputs produced. Listing 3.3 shows an excerpt of the states section of the ComputeActionThread thread of our ACC example.

```
states
  Waiting: initial complete state;
  Started, Detected, ..., ComputeBreak: state;
```

Listing 3.3 Sample of the states section of the behavior annex of the ComputeActionThread thread

3.5.2.4 Transitions Section

The transitions section defines transitions from a source state to a destination state. Transitions in a behavior automaton represent the execution sequence within a thread. A transition out of a complete state is initiated by a dispatch once the dispatch condi-tion is satisfied. Transitions can be guarded by dispatch conditions, or execute con-ditions, and can have actions. Listing 3.4 presents three transitions (t0, t1 and t3) from the transitions section of the behavior annex of the ComputeActionThread thread.

```
transitions
  --Periodically check input data
  t0 :
  Waiting -[on dispatch]-> Started {
    actual_speed:=inActualSpeed
  };
  --Detecting obstacles
  t1 :
  Started -[inObstacleDetected and (actual_speed != 0)]-> Detected {
    obstacle_distance:=inObstacleDistance
  };
...
  --Stopping car in case of emergency
  t3 :
  Emergency -[]-> SaveValues {
    outFullBreak!
  };
...
```

Listing 3.4 Sample of the transitions section of the behavior annex of the ComputeActionThread thread

Dispatch conditions explicitly specify dispatch trigger conditions out of a complete state. A dispatch condition is a Boolean expression that specifies the logical combination of triggering events: arrival of an event or event data on an event port or an event data port, receipt of a call on a provided subprogram access, or timeout event.

Execute conditions specify transition conditions out of an execution state to another state. They effectively select between multiple transitions out of a given state to different states. These conditions are logical expressions based on component inputs, subcomponent outputs, and values of data components, state variable values, and property constants. They can also result in catching a previously raised execution timeout exception.

If transitions have been assigned a priority number, then the priority determines the transition to be taken. The higher the priority number is, the higher the priority of the transition is. If more than one transition out of a state evaluates its condition to true and no priority is specified, then one transition is chosen non-deterministically. For multiple transitions with the same priority value the selection is also non-deterministic. Transitions with no specified priority have the lowest priority.

Each transition can have actions. Actions can be subprogram calls, retrieval of input and sending of output, assignments to variables, read/write to data components, and time consuming activities. An action is related to the transition and not to the states: if a transition is taken, the sequence of actions is performed and then the state specified as the destination of the transition becomes the new current state.

3.5.2.5 Transition Semantics

States of a behavior annex transition system can be either observable from the outside (initial, final or complete states), that is states in which the execution of the component is paused or stopped and its outputs are available; or non observable, execution states, that is internal states. The semantics of the AADL is concerned with the observable states of the automaton. The set S_A of automaton A thus only contains states corresponding to these observable states and set T_A big-step transitions from an observable state to another (by opposition with small-step transitions from or to an execution state).

A transition `behavior_transition` has source state s = `source_state_identifier`. Its guard formula g is defined by the translation of the expression `behavior_condition` as a logical formula. Its target state d = `destination_state_identifier` is that of the transition system defined by the semantic function $T(s, d)$ (defined Sect. 3.5.4) applied to its action block `behavior_action_block`.

A `transition_identifier`, if present, is represented by a label L that names the clock of the transition. It is a (virtual) event considered present and true if and only if the guard formula of that transition holds and the constraint of the automaton is enforced: the transition $(L : s, g, f, d)$ is equivalent to the transition (s, g, f, d) with the constraint $\hat{} L \Leftrightarrow (\hat{} s \wedge g)$.

A `behavior_transition_priority`, if present, enforces a deterministic logical order of evaluation among transitions. A pair of transitions $(s[m], g1, f1, s1)$ and $(s[n], g2, f2, s2)$ from a state s and such that $m < n$ (to mean that m has higher priority than n) is equivalent to the transitions $(s, g1, f1, s1)$ and $(s, g2 \wedge \neg g1, f2, s2)$: the guard formula of a prioritized transition is subtracted from all transition in the same state of lower or no priority.

```
behavior_transition ::=
  [ transition_identifier [ [ behavior_transition_priority ] ] : ]
  source_state_identifier { , source_state_identifier }*
    -[ behavior_condition ]-> destination_state_identifier
  [ behavior_action_block ] ;
```

3.5.3 Behavior Conditions

Behavior conditions that cause transitions may be either execute conditions or dispatch conditions.[4]

```
behavior_condition ::= execute_condition | dispatch_condition
```

Execute conditions are Boolean-valued expressions, and may only be used in transitions leaving an execution (or initial) state. State machines may never 'stall' in execution states; there must always be an enabled, outgoing transition from an execution state. The `otherwise` condition occurs when no other execute condition of a transition leaving an execution state is true.

---

[4]The grammar for `behavior_condition`, here, is slightly simplified from that in the BA standard.

```
execute_condition ::= logical_value_expression | otherwise
```

Dispatch conditions can only be associated with transitions from a `complete` state. A thread scheduler evaluates dispatch conditions to determine when threads are dispatched. A dispatch trigger condition can be the arrival of events or event data on ports (expressed as a disjunction of conjunctions) or timeout.

Periodic dispatches are always considered to be implicit unconditional dispatch triggers on complete states and handled by dispatch conditions without dispatch trigger condition. This is the case for transition `t0` presented in Listing 3.4.

```
dispatch_condition  ::= on dispatch [ dispatch_trigger_condition ]
      [ frozen ( frozen_ports ) ]

dispatch_trigger_condition ::= dispatch_trigger_logical_expression
    | stop | timeout_catch
```

Dispatch can be triggered by arrival of events at an `event port` or event-data at an `event data port`. To provide flexibility, dispatch conditions may be a disjunction, of conjunctions, of event (data) arrival at event (data) ports. Dispatch can also be triggered by event arrival at the predeclared `Stop` port.

Timeout catch is a dispatch trigger condition that is raised after the specified amount of time since the last dispatch or the last completion is expired.

```
timeout_catch ::= timeout
    [ [ ( port_identifier { or port_identifier }* ) ] behavior_time ]

dispatch_trigger_logical_expression ::=
    dispatch_conjunction { or dispatch_conjunction }*

dispatch_conjunction ::= port_identifier { and port_identifier }*
```

3.5.3.1 Behavior Condition Semantics

A `dispatch_condition` is represented by a guarding formula g that is formed by referring to the clock $\hat{\ }p$ of the logical combination of ports specified as its `dispatch_trigger_condition`.

An `execute_condition` is represented by a guarding formula that encodes its `logical_value_expression` using the current state of its persistent variables V. The `otherwise` clause is handled as the guard of least priority. The `otherwise` guard, if present in a transition leaving execution state s, applies if none of the guards from other transitions leaving s are true. It is hence defined by $(\hat{\ }s - tick_A(s))$, which differs from the stuttering clock of s, $\tau_A(s)$.

In the case of a time-triggered dispatch, when the dispatch trigger condition of an `on dispatch` clause is empty, the Boolean true is assumed, but only in the scope of the denoted object. It means that the dispatch condition is considered to be present as soon as it is triggered and an event is to be handled (otherwise, it can be regarded as silent, i.e., absent).

A `timeout` clause, if present, is denoted by the dispatch of the virtual event port `timeout`, whose trigger is associated with a real time constraint of the parent component behavior action block. *The parent component is responsible for triggering this event by respecting the real time constraint behavior time, if specified, as well as with the specified frozen ports list, if present.*

3.5.4 Action Language

The action language of BA defines actions performed during transitions. Actions associated with transitions are action blocks that are built from basic actions and a minimal set of control structures: sequences, sets, conditionals and loops. Action sequences are executed in order, while actions in actions sets can be executed in any order.

Basic actions can be assignment actions, communication actions or time consuming actions. Assignments consist of a value expression and a target reference (local variables, data components acting as persistent state variables, or outgoing features such as ports and parameters) for the value assignment, separated by the assignment symbol `:=`. For example transitions `t0` and `t1` presented in Listing 3.4 both have associated assignment actions.

Communication actions can be freezing the content of incoming ports, initiating a send on an event, data, or event data port, initiating a subprogram call or catching a previously raised execution timeout exception. Listing 3.4 presents the transition `t3` with associated action to initiate a send on the event port `outFullBreak`.

Timed actions can be predefined computation actions. Computation actions specify computation time intervals. An execution timeout exception can be raised after any behavior action block. Raising such a timeout event may trigger a transition with a timeout catch execute condition.

3.5.4.1 Action Semantics

Let us recall that the transition system T representing a behavior transition is defined by $T = (s, g, true, s') \bigcup T'$. It has source state s and a guard formula g. Its target state d is that of the transition system T' defined by the semantic function call encoding the *behavior_action_block* block as $\mathcal{T}(s, d)[behavior\_action\_block] = T'$. T' is constructed by recursively calling function $\mathcal{T}$ on the action block's sub-expressions.

```
behavior_action_block ::=   { behavior_actions } [ timeout behavior_time ]
behavior_actions ::=
  behavior_action | behavior_action_sequence | behavior_action_set
behavior_action_sequence ::=   behavior_action { ; behavior_action }+
behavior_action_set ::=   behavior_action { & behavior_action }+
behavior_action ::=
  basic_action | behavior_action_block
  | if ( logical_value_expression ) behavior_actions
    { elsif ( logical_value_expression ) behavior_actions }*
    [ else behavior_actions ]
    end if
  | for ( element_identifier in element_values ) { behavior_actions }
  | forall ( element_identifier in element_values ) { behavior_actions }
  | while ( logical_value_expression ) { behavior_actions }
  | do behavior_actions until ( logical_value_expression )
basic_action ::=   assignment_action | communication_action | timed_action
```

The recursive function $\mathcal{T}(s, d)[behavior\_actions] = T$ associates the action block *behavior_actions* guarded by a behavior condition of formula g, of source and target states s and d, to a transition system T. It is defined by case analysis on *behavior_actions*:

- a behavior action sequence is represented by concatenating the transition systems of its elements. For instance, $\mathcal{T}(s, d)[action_1 ; action_2]$ is represented by the union $T_1 \bigcup T_2$ of its transition systems $T_1 = \mathcal{T}(s, e)[action_1]$ and $T_2 = \mathcal{T}(e, d)[action_2]$, by introducing a new execution state e;
- a behavior action set is represented by composing the transition systems of its elements. For instance, $\mathcal{T}(s, d)[action_1 \& action_2]$ is represented by the synchronous composition

$$T = (T_1 | T_2)[(s_1, s_2)/s, (d_1, d_2)/d]$$

of its transition systems $T1 = \mathcal{T}(s_1, d_1)[action_1]$ and $T2 = \mathcal{T}(s_2, d_2)[action_2]$, substituting the composed states (s_1, s_2) and (d_1, d_2) by s and d.

A behavior action is defined by case analysis of its form:

- **if** (b) a_1 **else** a_2 **end if** is represented by a guard formula g corresponding to *logical_expression* and returning the union

$$T = T_1 \bigcup T_2 \bigcup \{(s, g, true, s_1), (s, \neg g, true, s_2)\}$$

of its transition systems $T_1 = \mathcal{T}(s_1, d)[a_1]$ and $T_2 = \mathcal{T}(s_2, d)[a_2]$ where the guard formula g is the translation of the logical value expression, b;
- **while** (b) a is represented by the union $T_1 \bigcup T_2$ of its transition systems $T_1 = \mathcal{T}(s_1, s_2)[a]$ and $T_2 = \{(s, h, true, s_1), (s, \neg h, true, d), (s_2, h, true, s_1), (s_2, \neg h, true, d)\}$ where the guard formula h is the translation of the logical value expression, b;
- **do** a **until** (b) is represented by the union $T_1 \bigcup T_2$ of its transition systems $T_1 = [a]$ and $T_2 = \{(s_1, h, true, s), (s_1, \neg h, true, d)\}$ where the guard formula h is the translation of the logical value expression, b;

- **forall** (j **in** e) a can be represented by the action set a_1 &...& a_n where a_i results from the substitution of j by the ith element value of e in a.
- **for** (j **in** e) a can be represented by the action sequence a_1 ; ...; a_n where a_i results from the substitution of j by the ith element value of e in a.

A basic action is defined by case analysis of its grammar's sub-clauses:

- an assignment action to a variable $v := e$ is represented by updating v with e as $\mathcal{T}(s, d)[v := e] = \{(s, true, v' = e, d)\}$ where v' represents the next value of v;
- an output port action *port!(value)* is represented by an action formula that binds *value* to *port* by $\mathcal{T}(s, d)[port!value] = \{(s, true, port = value, d)\}$;
- an input port action *port?(target)* is represented by an action formula that updates *target* to *port* by $\mathcal{T}[port?target] = \{(s, true, target' = port, d)\}$;
- a timed action of the form **computation** ($t_1[..t_2]$) is a timing constraint imposed on the execution time of the action block. It can either be represented by a timing property of the parent thread object or simulated by a protocol interacting with the scheduler using two virtual ports *ps* (start) and *pf* (finish) to specify a delay of time between exclusive occurrences of *ps* and *pf*, and to translate the timing specification by $\mathcal{T}(s, d)(t_1[..t_2)]) = \{(s, true, ps, c), (c, pf, true, d)\}$ using a complete state c and the timed constraint $@ps + t_1 \leq @pf + t_2$;
- subprogram invocations are specified using the communication protocols HSER, LSER or ASER (cf. Sect. 3.5.7). A subprogram invocation is hence represented by the composition of the client (the caller) and server (the callee) with the behavior of the calling protocol. For instance, a subprogram call *subprogram!(parameter)* using the HSER protocol is encoded by $\mathcal{T}(s, d)[subprogram!(parameter))] = \{(s, true, sps = pv, c), (c, spf, true, d)\}$. The output port *sps* encodes the call, the variable *pv* its parameter, and the input port *spf* signals the return from the callee;

3.5.5 Communication Actions

The communication actions defined by BA allow threads to interact with each other.

Threads can interact through shared data, connected ports and subprogram calls. The AADL execution model defines the way queued event/data of a port are transferred to the thread in order to be processed and when a component is dispatched.

Messages can be received by the annex subclause through declared features of the current component type. They can be in or in out data ports; in or in out event ports; in or in out event data ports and in or in out parameters of subprogram access.

The AADL standard defines that input on ports is determined by default *freeze* at dispatch time, or at a time specified by the *Input_Time* property and initiated by a *Receive_Input* service call in the source text. From that point in time the input of the port during this execution is not affected by arrival of new data, events, or event data until the next time input is frozen. For example, after transition t0 (in Listing 3.4) is fired by the periodic dispatch of the thread, all input ports of the thread are *frozen*, new arrival of data or events will not be taken into account before the next periodic dispatch.

The AADL standard also defines that data from data ports are made available through a port variable with the name of the port. The same transition t0 in Listing 3.4 uses the port variable inActualSpeed to get the data available on the same name port. If no new value is available since the previous freeze, the previous value remains available and the variable is marked as not fresh. Freshness can be tested in the application source code via service calls.

3.5.6 Expression Language

The expression language of BA is used to define expressions, the results of which are used either as logical conditions of transitions or conditional statements, or as values for assignment actions. Expressions consist of logical expressions, relational expressions, and arithmetic expressions. Values of expressions can be variables, constants or the result of another expression.

Variable expression values are evaluated from incoming ports and parameters, local variables, referenced data subcomponents, as well as port count, port fresh, and port dequeue. For example, transition t1 presented in Listing 3.4 is conditioned by an expression based on one event input (inObstacleDetected) and one variable value (actual_speed). Constant expression values are Boolean, numeric or string literals, property constants or property values.

3.5.7 Synchronization Protocols

Thanks to subprogram access features, an AADL thread can receive execution requests and execute the corresponding subprogram. With proper statements in a behavior annex subclause, it is possible to specify the states where specific requests can be accepted, which correspond to Ada selective accept statements or to HOOD[5] (Hierarchical Object-Oriented Design) functional activation conditions. This mechanism also allows a clean separation between the functional part of the component defined by a set of subprograms and the synchronization aspects specified by the behavior annex automaton. The internal behavior of a server component together with the specification of the interaction protocols between the server component and its clients define the global synchronization aspects.

The behavior annex introduces precise communication protocols that can be used to better control the blocking duration of a client thread during a remote call to a server thread. These protocols are derived from the main HOOD functional execution requests:

---

[5]http://www.esa.int/TEC/Software_engineering_and_standardisation/TECKLAUXBQE_0.html.

- HSER for Highly Synchronous Execution Request;
- LSER for Loosely Synchronous Execution Request;
- ASER for ASynchronous Execution Request.

3.5.7.1 Synchronization Semantics

Let cs and cd delimit the source and target state of subprogram call. Let ss and sd delimit the transition system of the server's subprogram. Let pc be the client request port and ps be the server reply port.

- The HSER protocol is encoded by the client transitions $\{(cs, true, pc, s), (s, \hat{}ps, true, cd)\}$, using a complete state s, and the server transition $\{(s_0, \hat{}pc, true, ss), (sd, true, ps, s_0)\}$;
- the LSER protocol is encoded by the client transitions $\{(cs, true, pc, s), (s, \hat{}ps, true, cd)\}$ and the server transition $\{(s_0, \hat{}pc, ps, ss)\}$;
- the ASER protocol is encoded by the client transitions $\{(cs, true, pc, s)\}$ and the server transition $\{(s_0, \hat{}pc, true, ss)\}$.

3.6 Related Work

Many related works have contributed to the formal specification, analysis and verification of AADL models and its annexes, hence implicitly or explicitly proposing a formal semantics of the AADL in the model of computation and communication of the verification framework considered.

The analysis language REAL [9] allows to define structural properties on AADL models that are checked inductively visiting the object of a model under verification. [8] presents an extension of this language called LUTE which further uses PSL (Property Specification Language) to check behavioral properties of models as well as a contract framework called AGREE for assume-guarantee reasoning between composed AADL model elements.

The COMPASS project has also proposed a framework for formal verification and validation of AADL models and its error annex [7]. It puts the emphasis on capturing multiple aspects of nominal and faulty, timed and hybrid behaviors of models. Formal verification is supported by the nuSMV tool. Similarly, the FIACRE framework [3] uses executable specifications and the TINA model checker to check structural and behavioral properties of AADL models.

RAMSES [6], on the other hand, presents the implementation of the AADL behavior annex. The behavior annex supports the specification of automata and sequences of actions to model the behavior of AADL programs and threads. Its implementation OSATE proceeds by model refinement and can be plugged in with Eclipse-compliant backend tools for analysis or verification. For instance, the RAMSES tools uses OSATE to generate C code for OSs complying the ARINC-653 standard.

Synchronous modeling is central in [16], which presents a formal real-time rewriting logic semantics for a behavioral subset of the AADL. This semantics can be directly executed in Real-Time Maude and provides a synchronous AADL simulator (as well as LTL model-checking). It is implemented by the tool AADL2MAUDE using OSATE.

Similarly, Yang et al. [19] defines a formal semantics for an implicitly synchronous subset of the AADL, which includes periodic threads and data port communications. Its operational semantics is formalized as a timed transition system. This framework is used to prove semantics preservation through model transformations from AADL models to the target verification formalism of timed abstract state machine (TASM).

Our proposal carries along the same goal and fundamental framework of the related work: to annex the core AADL with formal semantic frameworks to express executable behaviors and temporal properties, by taking advantage of model reduction possibilities offered thanks to a synchronous hypothesis, of close correspondence with the actual semantics of the AADL.

Yet, we aim at an effort to structure and use these concepts within the framework of a more expressive multi-rate or multi-clocked, synchronous, model of computation and communication: that of polychrony. Polychrony would allow us to gain abstraction from the direct specification of executable, synchronous, specification in the AADL, yet offer services to automate the synthesis of such, locally synchronous, executable specification, together with global asynchrony, when or where ever needed.

CCSL, the clock constraint specification language of the UML profile MARTE [15], relates very much to the effort carried out in the present Chapter. CCSL is an annotation framework to making explicit timing annotation to MARTE objects in an effort to disambiguate its semantic and possible variations.

CCSL actually provides a clock calculus of greater expressivity than polychrony, allowing for the expression of unbounded, asynchronous, causal properties between clocks (e.g. inf and sup).

While CCSL essentially is isolated as an annex of the MARTE standard for specifying annotations, our approach is instead to build upon the semantics of the existing behavior annex and specify it within a polychronous MoCC.

Finally, the Behavior Language for Embedded Systems with Software (BLESS) [10, 11] was derived from BA by adding non-executable assertions to behaviors. With human guidance, a proof engine transforms proof outlines into deductive proofs that every execution conforms to a formal behavior specification. Although the formal semantics defined for BLESS are expressed much differently than the semantics for BA defined here, they are not incompatible. We are endeavoring to merge the semantics so that deductively proved BLESS behaviors can also be analyzed with polychronous tools such as Polychrony.

Our previous work demonstrated that the all concepts and artifact of the AADL core could, as specified in its normative documents, be given an interpretation in the polychronous model of computation and communication [5, 13, 14, 20–22], by mean of its import and simulation in the Eclipse project POP's toolset.[6]

3.7 Conclusion

We propose a formal semantics for a significant subset of the behavioral specification annex of the Architecture Analysis and Design Language (AADL). This annex allows one to attach a behavior specification to any components of a system modeled using the AADL, and can be then analyzed for different purposes which could be, for example, the verification of logical, timing or scheduling requirements.

The addressed subset includes the transition system (state variables, states and transitions), the conditions that can be attached to transitions, the action language allowing to describe actions to be computed when a transition is fired and the expression language, used for logical conditions and assignment actions.

The semantics we presented for this subset relies on constrained automata (automata with variables derived from polychronous automata) and supports unambiguous reasoning, formal verification and simulation of the modeled system.

In future work, we will provide semantics for the remaining subset of the behavior specification annex of the AADL (mainly the synchronization protocols allowing to send and receive execution request in a client-server configuration). We will also implement the semantics of the behavior specification annex through a model transformation from the annex to the Signal language, in which the constrained automata are ahead implemented.

Acknowledgements This work was partly funded by Toyota InfoTechnology Center (ITC) and by INRIA D2T's standardisation support program. The authors wish to thank Pierre Dissaux, and all the SAE sub-committee on the AADL for valuable comments on the model and method presented in this work.

References

1. Aerospace Standard AS5506A: Architecture Analysis and Design Language (AADL) (2009)
2. Aerospace Standard AS5506/2: SAE Architecture Analysis and Design Language (AADL) Annex Volume 2, Annex D: Behavior Model Annex (2011)
3. B. Berthomieu, J.-P. Bodeveix, S. Dal Zilio, P. Dissaux, M. Filali, P. Gaufillet, S. Heim, F. Vernadat, Formal verification of AADL models with Fiacre and Tina, in *ERTSS 2010—Embedded Real-Time Software and Systems*, Toulouse, France, pp. 1–9, May 2010
4. L. Besnard, E. Borde, P. Dissaux, T. Gautier, P. Le Guernic, J.-P. Talpin, *Logically timed specifications in the AADL: a synchronous model of computation and communication (recommendations to the SAE committee on AADL)*. Technical Report RT-0446 (INRIA, April 2014)

---

[6]Polarsys Industry Working Group, Eclipse project POP, http://www.polarsys.org/projects/polarsys.pop.

5. L. Besnard, A. Bouakaz, T. Gautier, P. Le Guernic, Y. Ma, J.-P. Talpin, H. Yu, Timed behavioural modelling and affine scheduling of embedded software architectures in the AADL using Polychrony. Sci. Comput. Program., 54–77, August 2015
6. E. Borde, S. Rahmoun, F. Cadoret, L. Pautet, F. Singhoff, P. Dissaux, Architecture models refinement for fine grain timing analysis of embedded systems, in *25th IEEE International Symposium on Rapid System Prototyping, RSP 2014*, New Delhi, India, 16–17 Oct 2014
7. M. Bozzano, R. Cavada, A. Cimatti, J.-P. Katoen, V. Yen Nguyen, T. Noll, X. Olive, Formal verification and validation of AADL models, in *Proceedings of Embedded Real Time Software and Systems Conference*, 2010
8. D. Cofer, A. Gacek, S. Miller, M.W. Whalen, B. LaValley, L. Sha, Compositional verification of architectural models, in *Proceedings of the 4th International Conference on NASA Formal Methods, NFM 2012* (Springer, Berlin, 2012), pp. 126–140
9. O. Gilles, J. Hugues, Expressing and enforcing user-defined constraints of AADL models, in *2014 19th International Conference on Engineering of Complex Computer Systems*, pp. 337–342, 2010
10. B.R. Larson, P. Chalin, J. Hatcliff, BLESS: formal specification and verification of behaviors for embedded systems with software, in *Proceedings of the 2013 NASA Formal Methods Conference*. Lecture Notes in Computer Science, vol. 7871 (Springer, Heidelberg, 2013), pp. 276–290
11. B.R. Larson, Y. Zhang, S.C. Barrett, J. Hatcliff, P.L. Jones, Enabling safe interoperation by medical device virtual integration, in *IEEE Design and Test*, Oct 2015
12. P. Le Guernic, T. Gautier, J.-P. Talpin, L. Besnard, Polychronous automata, in *TASE 2015, 9th International Symposium on Theoretical Aspects of Software Engineering*, Nanjing, China, IEEE Computer Society, Sept 2015, pp. 95–102
13. Y. Ma, H. Yu, T. Gautier, P. Le Guernic, J.-P. Talpin, L. Besnard, M. Heitz, Toward polychronous analysis and validation for timed software architectures in AADL, in *The Design, Automation, and Test in Europe (DATE) Conference*, Grenoble, France, 2013, pp. 1173–1178
14. Y. Ma, H. Yu, T. Gautier, J.-P. Talpin, L. Besnard, P. Le Guernic, System synthesis from AADL using polychrony, in *Electronic System Level Synthesis Conference*, June 2011
15. F. Mallet, J. DeAntoni, C. André, R. de Simone, The clock constraint specification language for building timed causality models. Innov. Syst. Softw. Eng. **6**(1), 99–106 (2010)
16. P.C. Ölveczky, A. Boronat, J. Meseguer, Formal semantics and analysis of behavioral AADL models in Real-Time Maude, in *Proceedings of the 12th IFIP WG 6.1 International Conference and 30th IFIP WG 6.1 International Conference on Formal Techniques for Distributed Systems*, FMOODS 2010/FORTE2010 (Springer, Berlin, 2010), pp. 47–62
17. B. Selic, S. Gérard, *Modeling and Analysis of Real-Time and Embedded Systems with UML and MARTE: Developing Cyber-Physical Systems* (Morgan Kaufmann Publishers Inc., San Francisco, CA, USA, 2013)
18. M. Skoldstam, K. Akesson, M. Fabian, Modeling of discrete event systems using finite automata with variables, in *46th IEEE Conference on Decision and Control*, pp. 3387–3392 (2007)
19. Z. Yang, K. Hu, J.-P. Bodeveix, L. Pi, D. Ma, J.-P. Talpin, Two formal semantics of a subset of the AADL, in *16th IEEE International Conference on Engineering of Complex Computer Systems, ICECCS 2011*, Las Vegas, Nevada, USA, 27–29 April 2011, pp. 344–349, 2011
20. H. Yu, Y. Ma, T. Gautier, L. Besnard, P. Le Guernic, J.-P. Talpin, Polychronous modeling, analysis, verification and simulation for timed software architectures. J. Syst. Archit. **59**(10), 1157–1170 (2013)
21. H. Yu, Y. Ma, T. Gautier, L. Besnard, J.-P. Talpin, P. Le Guernic, Y. Sorel, Exploring system architectures in AADL via Polychrony and SynDEx. Front. Comput. Sci. **7**(5), 627–649 (2013)
22. H. Yu, Y. Ma, Y. Glouche, J.-P. Talpin, L. Besnard, T. Gautier, P. Le Guernic, A. Toom, O. Laurent, System-level co-simulation of integrated avionics using Polychrony, in *ACM Symposium on Applied Computing*, TaiChung, Taiwan, March 2011, pp. 354–359

Chapter 4
MARTE for CPS and CPSoS

Present and Future, Methodology and Tools

Frédéric Mallet, Eugenio Villar and Fernando Herrera

Abstract Cyber-Physical Systems (CPS) combine discrete computing elements together with physical devices in uncertain environment conditions. There have been many models to capture different aspects of CPS. However, to deal with the increasing complexity of these ubiquitous systems, which invade all the part of our lives, we need an integrated framework able to capture all the different views of such complex systems in a consistent way. We also need to combine tools to analyze their expected properties and guarantee safety issues. Far from handing out a full-fledge solution, we merely explore a possible path that could bring part of the solution. We advocate for relying on UML models as a *unifying* framework to build a single-source modeling environment with design, exploration and analysis tools. We comment on some useful extensions of UML, including MARTE and SysML, and show how they can together capture different views of CPS. We also report on some recent results obtained and discuss possible evolutions in a near future.

Keywords MARTE · Cyber-Physical Systems · Systems of systems · Model-driven methodology · System engineering

4.1 Introduction

4.1.1 CPS and CPSoS

Cyber-Physical Systems combine digital computational systems with surrounding physical processes. Computations are meant to control and monitor the physical environment, which in turn affects the computations.

F. Mallet (✉)
Université Côte d'Azur, CNRS, Inria, I3S, 06900 Sophia Antipolis, France
e-mail: Frederic.Mallet@unice.fr

E. Villar · F. Herrera
TEISA, GESE, Universidad de Cantabria, Av. Castros sn, 39011 Santander, Spain
e-mail: evillar@teisa.unican.es

F. Herrera
e-mail: fherrera@teisa.unican.es

© Springer Nature Singapore Pte Ltd. 2017
S. Nakajima et al. (eds.), *Cyber-Physical System Design from an Architecture Analysis Viewpoint*, DOI 10.1007/978-981-10-4436-6_4

The main characteristics of Cyber Physical Systems and main design challenges have been identified some years ago [22, 23]. CPS are: heterogeneous, in the sense that they combine various models of computations relying on both discrete and continuous time abstractions; platform-aware and resource-constrained, and thus the software depends on various non-functional properties imposed by the platform; time-sensitive and often safety-critical; widely distributed with heterogeneous interconnects.

CPS are first and foremost complex systems and as such designing them requires several models, usually hierarchical, to fully capture the different aspects and views, whether structural or behavioral. Structural models include a description of the components or blocks of the systems and of the communication media involved. Behavioral models include hierarchical state machines and dataflow streaming diagrams. Expected or faulty interactions with the surrounding environment can be captured as a set of use cases or requirements that correspond to positive or negative scenarios. Such models are usually called **heterogeneous** in the sense that they combine different models, each of which may follow a different *model of computation*.

CPS also have the main characteristics of embedded systems, which are usually **platform-aware**. Contrary to standard software engineering, embedded system design depends a lot on the execution platform on which the system should execute, be it a system-on-a-chip (SoC), with multiple computing resources and a complex memory hierarchy, or a wide scale distributed system, with potentially all the variety of interconnects and communication media. This awareness of the platform makes it important to account for how and when the available resources are accessed or 'consumed', considering together both the spatial and temporal dimensions. The spatial dimension is not only about how much resource is available but also about where the resources are physically located in the system relative to each other. How much resource is available is indeed easy since it is usually given by the technology used and the targeted selling price. However, how the resources are used makes all the difference between two a priori equivalent products. The spatial dimension encompasses the interconnect topology, i.e., physical parallelism available, but also and more importantly where the data and programs are allocated. Indeed, the distance between the data memory and the computing resource that executes the program largely impact the fetching time that may potentially largely exceed the computing time. Then, this spatial distribution requires to perform the temporal scheduling of both the execution of programs and the routing of data from memory to computing resources, forth and back. This leads to logical concurrency coming both from the physical parallelism and the inherent data and control dependencies of the application. CPS are therefore **resource-constrained** and **time-sensitive** systems. Even though the resources (memory size, computing power) are not necessarily as scarce as they used to be, nevertheless finding the right tradeoff between the resource usage, the computation speed and the cost makes it a multi-criteria optimization problem difficult to solve. The cost is not only measured in terms of money, but includes all kinds of additional extra functional properties (like power, energy, thermal dissipation), also called **non-functional properties**.

More than being mere embedded systems, CPS are usually made of multiple inter-connected embedded subsystems, some of which are computing devices and some other being physical devices. This requires some abilities to describe **heterogeneous interconnects**, while simpler embedded systems usually only rely on homogeneous communication structures. Being made of several computing devices also consti-tute a big step since it requires to model the whole system as a closed model, with software, devices but also with the environment and the expected **continuous inter-actions** with this environment. In standard software development, the environment is by definition outside the system to be developed. **Close loop systems** have been modeled for several years with tools and techniques to find approximate solutions to differential equations and are well established. However, the integration with dis-crete models still causes problems in many tools and each of them proposes ad-hoc solutions. A seamless integration with a generic environment is still to be proposed.

Finally, cyber-physical systems are often big and often interact directly with users that are not even aware of the computer. The size is an aggravating factor since a single system concerns potentially millions of people (smart cities, intelligent transportation systems…). It means some CPS are **safety-critical**, just like embedded systems but at a larger scale. The large scale and the integration of multiple systems gives rise to the notion of System of Systems. It also increases the demand to have sound and scalable models along with verification tools. Sometimes, they also require certification tools to be accredited and allowed to be used in public environments (e.g., unmanned aerial vehicle). However, we do not address at all the certification issue in this chapter.

Moore's Law has dominated the (re-)evolution of electronics during the last quar-ter of the 20th century. All the electronic products we use today depend directly or indirectly on the increasing integration capability allowed by semiconductor technol-ogy. If Moore's Law has changed the world, its end may have a similar effect. For the first time, the underlying technology will be stable with only incremental improve-ments in time. Cyber-Physical Systems of Systems (CPSoS [9])[1] will dominate the electronics century becoming pervasive in all the aspects of our daily lives. In this new scenario, modeling, analysis and verification of CPSs must evolve. The tendency is toward complex, heterogeneous, distributed networks of many computing nodes. Services will be offered by the interaction of functional components deployed onto many distributed computing resources of many kind, from small motes,[2] embedded systems and smart-phones to large data centers and even High-Performance Com-puting (HPC) facilities. Electronic design in this new context should address effec-tively new requirements. Among them, scalability, reusability, human interaction, easy modeling, fast design-space exploration and optimization, powerful functional and extra-functional verification, efficient handling of mixed-criticality and security. An essential aspect will be the availability of powerful, CPS modeling and analysis frameworks able to produce automatically efficient implementations of the system

---

[1] We use CPS or CPSoS interchangeably, while most considered systems are complex enough to be seen as an integration of systems with more or less explicit interactions.

[2] Motes: embedded devices consisting of sensors, radios, and microprocessors.

model on many different computing resources. The Unified Modeling Language (UML) is a very good candidate to provide the modeling means in this new context.

4.1.2 Role of UML and Its Extensions

The Unified Modeling Language [33] is a general-purpose modeling language specified by the Object Management Group (OMG). It proposes graphical notations to represent all aspects of a system from the early requirements to the deployment of software components, including design and analysis phases, structural and behavioral aspects. As a general-purpose language, it does not focus on a specific domain and relies on a weak, informal semantics to widen its application field. However, when targeting a specific application domain and especially when building trustworthy software components or for critical systems where lives may be at stake, it becomes necessary to extend the UML and attach a formal semantics to its model elements. The simplest and most efficient extension mechanism provided by the UML is through the definition of profiles. A UML profile adapts the UML to a specific domain by adding new concepts, modifying existing ones and defining a new visual representation for others. Each modification is done through the definition of annotations (called stereotypes) that introduce domain-specific terminology and provide additional semantics. However, the semantics of stereotypes must be compatible with the original semantics (if any) of the modified or extended concepts, i.e., the base metaclass.

Domain-specific profiles, such as the UML Profile for Modeling and Analysis of Real-Time and Embedded systems (MARTE [32]) extends the UML with generic features required for real-time and embedded systems. Still, MARTE should be extended [34] to cover the modeling of the complete CPS and in particular the physical models that are usually left outside the scope of classical digital (cyber) models.

SysML [31] is another extension dedicated to systems engineering and has been successfully used to cope with physical models.

To summarize, CPS demand the integration of continuous models, classical state-based or dataflow models, hardware descriptions, and non-functional constraints. UML offers a tool-neutral non-proprietary solution that already contains most of the required notations. However, those notations need to be tailored to capture specific aspects of CPS (time, non-functional properties, continuous models). Both MARTE and SysML offer some extensions dedicated to these goals, and we discuss here some examples of useful features of either MARTE or SysML to model CPS. These notations also need to come with adequate, not tool-specific, explicit semantics if we are to address safety-critical issues.

4.1.3 Outline

Focusing on the key characteristics of CPS, we present some selected aspects of MARTE in Sect. 4.2. In Sect. 4.3, we select a simple case study addressed in a recent project and show which aspects can be modeled using UML, MARTE, SysML and some extensions that we thought were lacking in MARTE. More specifically, we explore two extensions, one to allow for the modeling of systems with mixed-criticality issues, another to deal with design space exploration. Section 4.4 describes some developed tools. Section 4.5 discusses the potential future use of UML in the modeling of CPSs, before our partial conclusion in Sect. 4.6.

4.2 Overview of MARTE

4.2.1 Overview

The UML profile for MARTE [32] extends the UML with concepts related to the domain of real-time and embedded systems. It supersedes the UML profile for Schedulability, Performance and Time (SPT [30]) that was extending the UML 1.x and that had limited capabilities. UML 2.0 has introduced a simple (or even simplistic) model of time and has proposed several new extensions that made SPT unusable. Therefore MARTE has been defined to be compatible with UML Simple Time model and now supersedes SPT as the official OMG specification.

SysML [13] is another extension dedicated to systems engineering. We use some notations from SysML and we introduce those notations when required. The task forces of MARTE and SysML have synchronized their effort to allow for a joint use of both profiles. The remainder of this subsection gives an overview of MARTE, which consists of three parts: Foundations, Design and Analysis. We try to give a general overview of most of it while focusing on the aspects that we use in our examples.

4.2.2 Foundations

The foundation part of MARTE is itself divided into five chapters: *CoreElements*, *NonFunctionalProperties* (NFP), *Time*, *GenericResourceModeling* (GRM) and *Allocation*.

4.2.2.1 CoreElements

Define configurations and modes, which are key parameters for analysis. However, their realization is mainly inspired from classical UML State Machines and we do not use specific constructs from this chapter.

4.2.2.2 NFP

Gives a support to describe Non-Functional properties. In real-time systems, pre-serving the non-functional (or extra-functional) properties (power consumption, area, financial cost, time budget…) is often as important as preserving the functional ones. The UML proposes no mechanism at all to deal with non-functional properties and relies on mere strings of characters for that purpose, while one could expect a richer type-based system to guarantee the well-formedness of expressions.

4.2.2.3 Time

Is often considered as an extra-functional property that comes as a mere annotation after the design. These annotations are fed into analysis tools that check the confor-mity without any actual impact on the functional model: e.g., whether a deadline is met, whether the end-to-end latency is within the expected range. Sometimes though, time can also be of a functional nature and has a direct impact on what is done and not only when it is done. All these aspects are addressed in the time chapter of MARTE.

4.2.2.4 GRM

Chapter provides annotations to capture the available resources on which the applica-tive part shall be deployed.

4.2.2.5 Allocation

Chapter gives a SysML-compatible way to make deployments. In MARTE, we use the wording allocation since the UML deployment usually implies (in people's mind) a physical distribution of a software artifact onto a physical node. Allocation in MARTE goes further. It encompasses the physical distribution of software onto hardware, but also of tasks onto operating system processes, and, more importantly, it covers the temporal distribution (or scheduling) of operating parts that need to share a common resource (e.g., several tasks executing on a single core processor, distributed computations communicating through an interconnect).

4.2.3 Non-functional Properties

The *NFP subprofile* offers mechanisms to describe the quantitative as well as the qual-itative aspects of properties and to attach a unit and a dimension to quantities. It defines a set of predefined quantities, units and dimensions and supports customization. NFP comes with a companion language called Value Specification Language (VSL) that defines the concrete syntax to be used in expressions of non-functional properties. VSL also recommends syntax for user-defined properties.

There are two levels of support in MARTE. The first level is a set of ready-to-use types that should be imported by users whenever relevant. For instance, the *MeasurementUnits* Library in MARTE provides a set of predefined units and dimensions, most of the ones that are relevant from the International System of Units (SI) [36]. Figure 4.1 (upper part) shows an except of such elements. A symbol is assigned to each dimension. This symbol is used to build new dimensions like *PowerUnitKind* defined as the square of length times a weight divided by a time to the power 3. The second level is a set of mechanisms to create constructs that do not exist in MARTE, for instance to build new units or new dimensions. Figure 4.1 (lower part) shows, on the left side, the creation of a new dimension to represent a Torque (or moment force), and, on the right-hand side, the creation of one new type. NFP_Weight is proposed as part of MARTE libraries and NFP_Torque is created following the same construction mechanisms. We shall use this dimension in subsequent examples to represent the torque of a rotor in a quadrotor system.

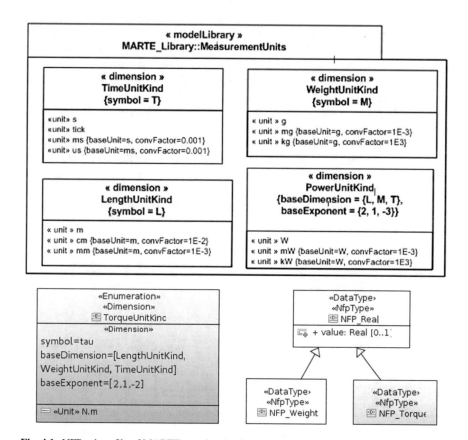

Fig. 4.1 NFP subprofile of MARTE: two levels of support

MARTE tools rely on such mechanisms to perform dimension analyses, which is of paramount importance for system engineering.

4.2.4 Time

The time model of MARTE has been extensively described before [1, 2, 26], we give here the minimum to understand its role for modeling CPS.

Time in SPT is a metric time with implicit reference to physical time. As a successor of SPT, MARTE supports this model of time. UML 2, issued after SPT, has introduced a model of time called *SimpleTime*. This model also makes implicit references to physical time, but is too simple for use in real-time applications, and was initially devised to be extended in dedicated profiles.

MARTE goes beyond SPT and UML 2. It adopts a more general time model suitable for system design. In MARTE, Time can be physical, and considered as continuous or discretized, but it can also be logical, and related to user-defined clocks. Time may even be multiform [4], allowing the use of different units (seconds, steps, processor ticks, heartbeats) to refer to temporal phenomena and allowing different time logics with progresses in a non-uniform fashion, and possibly independently to any (direct) reference to physical time. In MARTE, time is represented by a collection of *Clocks*. The use of word *Clock* comes from vocabulary used in the synchronous languages. They may be understood as a specific kind of events on which constraints (temporal, hence the name, but also logical ones) can be applied. Each clock specifies a totally ordered set of instants, i.e., a sequence of event occurrences. There may be dependence relationships between the various occurrences of different events. Thus this model, called the MARTE time structure, is akin to the Tagged Systems [21]. To cover continuous and discrete times, the set of instants associated with a clock can either be dense or discrete.

Figure 4.2 shows the main stereotypes introduced by MARTE Time subprofile.

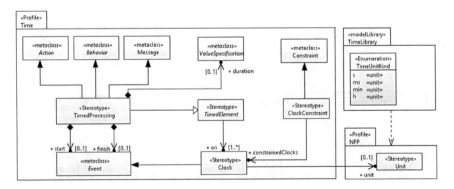

Fig. 4.2 Excerpt of MARTE time subprofile

Stereotype Clock is one important stereotype that extends UML metaclass Event. A Clock carries specific information such as its actual unit, and values of quantitative (resolution, offset...) or qualitative (time standard) properties, if relevant.

TimedElement is another stereotype introduced in MARTE. A timed element is an abstract stereotype that associates at least one clock with a modeling element. TimedProcessing is a specialization of TimedElement, which extends the UML meta-classes Action, Behavior and Message. It defines a start and a finish event for a given action/behavior/message. These events (which are usually clocks) specify when the action starts or when it finishes. TimedProcessing also specifies the duration of an action. Duration is measured on a given logical or physical clock. In a MARTE model of a system, stereotype TimedElement or one of its specializations is applied to model elements which have an influence on the specification of the temporal behavior of this system. The expected behavior of such TimedElements is controlled by a set of ClockConstraints. Those constraints specify dependencies between the various occurrences of events. Dedicated languages, like the Clock Constraint Specification Language (CCSL [25]) can be used to specify those constraints formally.

The MARTE Time subprofile also provides a model library named TimeLibrary. This model library defines the enumeration TimeUnitKind which is the standard type of time units for chronometric clocks. This enumeration contains units like s (second), its submultiples, and other related units (e.g., minute, hour). The library also predefines a clock called IdealClock, which is a dense chronometric clock with the second as time unit. This clock is assumed to be an ideal clock, perfectly reflecting the evolutions of physical time. It should be imported in user's models with references to physical time concepts (e.g., frequency, physical duration).

4.2.5 Allocation

Since embedded systems are platform-aware, one needs a way to map the elements of the application onto the execution platform. This aspect is specifically addressed by the allocation subprofile of MARTE, which is further described in this subsection.

The wording *Allocation* has been retained to distinguish this notion from UML Deployment diagrams. Deployments are reserved to deploy artifacts (e.g., source code, documents, executable, database table) onto deployment targets (e.g., processor, server, database system). The MARTE allocation is much more general than that. For instance, it is meant to represent the allocation of a program onto a system thread, or of a process onto a processor core. More generally, it is used to represent the association of an element (action, message, algorithm) that consumes a resource onto the consumed resource (processing unit, communication media, memory). Wordings 'mapping' or 'map' have also been discarded since they often refer to a function and then map one input from a domain to one single output in the co-domain. The allocation process, however, is an n-to-m association, take for instance a bunch of tasks that need to be scheduled on several cores.

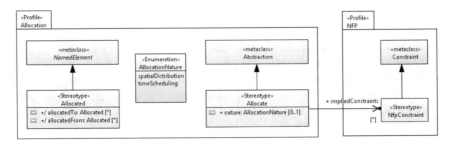

Fig. 4.3 Exerpt of MARTE allocation subprofile

Note that the wording *execution platform* has been preferred to 'architecture' or 'hardware'. Indeed, architecture is a way to describe the structure of a system, while an execution platform contains both structural and behavioral parts. On the other hand, the execution platform is not necessarily a piece of hardware. It can be a piece of software, a virtual machine, a middleware, an operating system or a mixed platform that combines software and hardware intellectual properties (IPs).

Finally, it is also important to note that this notion of allocation is common between MARTE and SysML in a bid to ease the combination of the two profiles. In particular, for CPS both profiles can and shall be used jointly [34].

Figure 4.3 shows the two main stereotypes of the subprofile, Allocate and Allocated. Allocate represents the allocation itself, while Allocated may be used on both sides to mark either the element that is allocated or the resource onto which an element is allocated. The property nature is meant to distinguish two kinds of possible allocations: *spatial* and *temporal*. Typically, when messages are allocated onto a buffer or a memory, this is a spatial allocation. Indeed, the message will consume/use some cells of the memory. However, when two tasks are allocated onto a processing unit, this is a temporal allocation (scheduling); It means those two tasks must be scheduled to avoid resource conflicts. When a program is allocated onto a processor, this can be considered both as spatial and temporal allocations; Spatial because the program consumes disk and memory resources; Temporal because while this program executes, another one cannot execute simultaneously. The allocation usually implies constraints that describe precisely the impact (or cost) of the allocation on the non-functional properties. This is why there is an association to a specific MARTE stereotype called NfpConstraint, i.e., to capture the constraints implied by the allocation in terms of memory consumption, power consumption, or execution time, for instance.

4.2.6   Design and Analysis in MARTE

4.2.6.1   The Design Part

Has four chapters: High Level Application Modeling (HLAM), Generic Component Modeling (GCM), Software Resource Modeling (SRM), and Hardware Resource Modeling (HRM).

The first chapter (HLAM) describes real-time units and active objects. Active objects depart from passive ones by their ability to send spontaneous messages or signals, and react to event occurrences. Normal objects, the passive ones, can only answer to the messages they receive or react on event occurrences. HLAM also introduces the notions of real-time feature and specification. A real-time feature («RtFeature») can be anything on which one wishes to pose some kinds of real-time constraints. The constraints themselves are applied via a real-time specification («RtSpecification»). We give some examples of such usage in Sect. 4.3.

While clocks are precisely sequences of time points where events repetitively occur, a real-time feature makes no assumption on the specific constraint to be applied, i.e., it can refer to time point or time intervals or something much more complex. However, as we show later, some phenomena can be described either as clocks and clock constraints or with real-time features and real-time specifications. We consider that the specification is a compact way to assign several kinds of properties to a generic feature while the semantics of such properties can be described at a finer grain using clocks and clock constraints.

The three other chapters provide a support to describe resources used and in particular execution platforms on which applications may run. A generic description of resources is provided, including stereotypes to describe communication media, storage and computing resources. Then this generic model is refined to describe software and hardware resources along with their non-functional properties.

4.2.6.2 The Analysis Part

Also has a chapter that defines generic elements to perform model-driven analysis on real-time and embedded systems.

This generic chapter is further specialized to address schedulability analysis and performance analysis.

The chapter on schedulability analysis is not specific to a given technique and addresses various formalisms like the classic and generalized Rate Monotonic Analysis (RMA), holistic techniques, or extended timed automata. This chapter provides all the keywords usually required for such analyzes.

Finally, the chapter on performance analysis, even if somewhat independent of a specific analysis technique, emphasizes concepts supported by the queuing theory.

The single-source modeling methodology introduced in Sect. 4.3 gives a way to combine all these mechanisms. However, there is no space here to describe them in length and we rather give methodological information in the following section.

4.3 MARTE for CPS

4.3.1 Case Study: Quadcopter

We consider the example of a quadcopter as an example of cyber-physical system. An efficient design of the flight control part requires coupling to the analysis a model of the dynamics of the quadcopter and of its environment (i.e., physical laws, data received by sensors).

Figure 4.5 gives a UML model of a quadcopter and combines several models to capture different views. The structural view (at the bottom right) is a UML composite structure model. We use MARTE allocation with its two acceptions. One structural allocation to allocate the PIDController on the Zinq board. It is identified as a spatial distribution. We also allocate the QuadRotorController to the PIDController as a time scheduling allocation. Indeed, all the actions have to be scheduled on the controller.

Looking at behavioral views, we use a UML activity to model the controller and its five main concurrent actions. All these actions have to compete for the available resources. «TimedProcessing» refines the description to assign a *clock* to the start instant of three actions. Two of them become synchronized, both start (synchronously) on clock $c2$. The last one starts with clock $c100$ and we use a clock constraint to denote the relative speed. Indeed $c100$ is fifty times slower than $c2$. Later (see Sect. 4.3.2), we can refine such a specification to assign a precise physical periodic behavior to each action.

Another behavioral view describes the operating modes, three while being in-flight. CoreElements stereotypes of MARTE serve to identify the modes. Switching from OnGround to InFlight is done according to events *on* and *off*. Both timeEvent become *clock* and a clock constraint explains their relative behavior.

Each of the three inflight-mode behaves differently. Here we describe one of the modes using SysML parametrics, which are acausal models introduced by SysML. The word acausal is used here in opposition to causal, i.e., there is no causal relationship among the variables of a formula, no assumption on which ones are inputs and which ones are outputs.

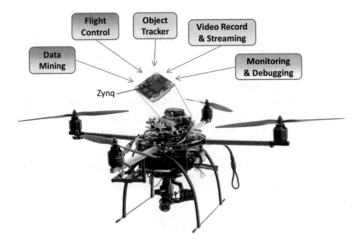

Fig. 4.4 Quadcopter system

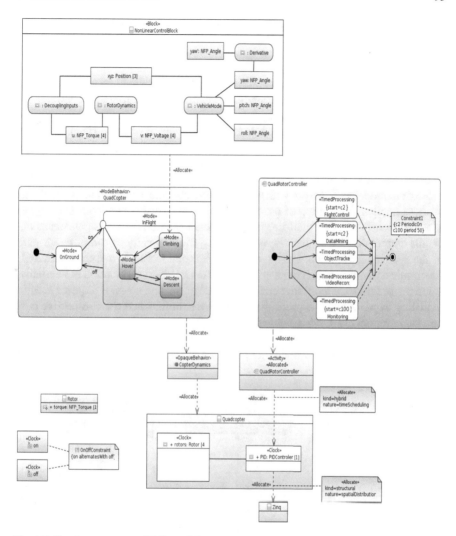

Fig. 4.5 Quadcopter system: UML model

The rectangles denote values or properties connected the pins, some of them use the NFP types defined earlier (like *NFP_Torque*). The rounded rectangles denote constraint blocks and define a non-linear equations among the values. This model clearly defines a closed loop to control the position (x, y, z) of the quadcopter. The content of the constraint blocks is typically defined as an equation that can feed a simulation in tools like Simulink.

4.3.2 Proposed Extensions: Mixed-Criticality, Design-Space Exploration

The growing capabilities and cost effective solutions provided by embedded platforms has increased their application domains. This can be illustrated through the Quadcopter system sketched in Fig. 4.4.

Moreover, the quadcopter system is also an example of Mixed-Criticality system (MCS) [7, 24]. The application software of this system has parts (data mining, flight control) which concern safety, while other parts concern the mission (e.g., object tracking, stream server) or even less important tasks (e.g., data logging). It shall be possible to model and analyze with sufficient accuracy the behavior and performance of these applications on top of a multi-core, costly effective platform. The capture of a sufficiently detailed platform model exposing shared resources, attributes and details concerning performances, and labeling the criticality of the different parts of the systems is required. Design space exploration (DSE) is a crucial activity in Electronic System Level (ESL) design [3]. It consists in the evaluation of alternative solutions to find a tradeoff between the performances and costs. It requires specific modeling capabilities to capture the alternative solutions and their respective cost, i.e., the potential available *design space*.

MARTE offers several mechanisms to cover most of these needs, as shown in the single-source modeling methodology proposed in [16, 38], and sketched in Fig. 4.6.

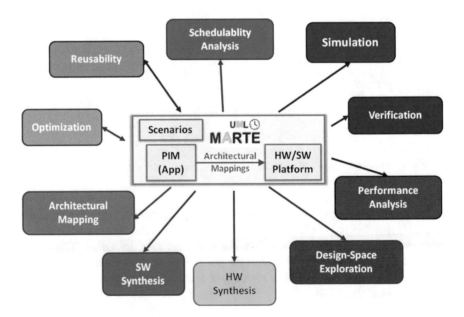

Fig. 4.6 Single-source analysis and design methodology of the University of Cantabria

The main idea is to rely on a common, unique system model, essentially UML and MARTE. Around this model, all the tool infrastructure enables a number of system-level design activities, which include verification, schedulability analysis, simulation, performance analysis, and software synthesis. In [16], the advantages and other features of the single-source methodology are discussed in detail. In the framework of the CONTREX project [29], the single-source design framework is implemented (see details in Sect. 4.4.2). The single-source modeling methodology enables to build up a component-based application model that can be mapped onto a platform. The platform model describes in detail the available HW and SW resources. It includes efficient multi-core platforms with shared resources, a main aspect to deal with in MCS design. The modeling methodology also states how to model a design space by relying on MARTE and on the Value Specification Language (VSL).

The single-source modeling methodology provides fixes to patch the minor deficiencies of the current MARTE specification [32]. They have to do with modeling of MCS and for DSE, which is understandable, as the standard was not initially designed for such purposes.

4.3.2.1 Extensions for Design Space Exploration

A basic modeling construct for DSE is the DSE parameter. A DSE parameter states that the value of an attribute associated with a modeling element can adopt one among a set of values during the DSE activity. Along the DSE, several solutions are assessed. A solution is defined by the set of values assigned to each of the DSE parameters. Figure 4.7 illustrates the modeling construct employed for describing a DSE parameter in the single-source methodology. A MARTE NFP constraint plus the *ExpressionContext* stereotype is linked to the model element (a HW resource component in Fig. 4.7) containing the attribute. The expression is captured in MARTE VSL with the following syntax:

$DSEParameterName = DSERangeSpecification$

The "$\$$" symbol prefixes a VSL variable, and thus the DSE parameter name. The *DSERangeSpecification* expresses the range of the DSE parameter, that is, all the values that the *DSEParameterName* variable can have during the exploration. The DSE parameter range can be annotated either as a collection or as an interval. Collections are captured with the syntax *DSERangeSpecification* = (v1, v2, v3, unit), where "v1, v2, v3" are the numerical values of the parameter and *unit* expresses the physical unit associated with the values. MARTE provides a rich set of unit kinds (see Sect. 4.2.3), to support the different extra-functional properties characterising systems components, e.g. frequencies, bandwidth, data size. Intervals follow the syntax "*DSERangeSpecification* = ($[v_{min}..v_{max}]$, unit)". For a complete determination of the exploration range, this style obliges to assume an implicit step. For an explicit and complete determination of the DSE range, the support of the style "*DSERangeSpecification* = ($[v_{min}..v_{max}]$, step], unit)" is proposed, which means a minor extension of VSL. The definition of non-linear ranges is possible. For instance, *step* can take

Fig. 4.7 DSE parameter
associated with the
frequency of a HW processor
component

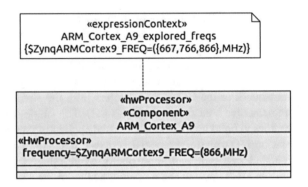

the value *exp2*, which enables the definition of a geometrical progression, i.e., the second value is "$v_{min}x2$", and so on.

Figure 4.7 illustrates the specification of a design space on the frequencies of the ARM processing cores of the quadcopter HW platform. This way, the exploration of the impact on performance depending on the selection of a Z-7020 device (which works at 667 MHz), a Z-7015 device (at 766 MHz) or a Z-7020 device (at 866 MHz) can be explored.

In Fig. 4.7, the DSE has been associated with the processor component declaration (in the HW platform resources view). Therefore, once the DSE parameter value is fixed, it is fixed for all its instances. The single-source methodology also allows the association of the DSE parameter with the instance properties. Moreover, the methodology enables the definition of DSE parameters at the application level, and the specification of parameterizable mappings. Further details can be found in [16, 38]. In any cases, as shown, MARTE (and specifically VSL) is exploited for most of the required constructs for DSE, while a minor extension is required (for explicitly stating the steps of an interval-based specification of a DSE parameter).

4.3.2.2 Extensions for Mixed-Criticality

The single-source modeling methodology [16, 38] was specifically extended in CONTREX [29] for supporting the MCS modeling and design. This requires a minor extension of MARTE, sketched in Fig. 4.8, and proposed to the OMG. The first extension adds a criticality attribute to a NFP constraint (Fig. 4.8, left hand side). The criticality attribute is an integer to denote an abstract criticality level. The NFP constraint can be associated with different types of modeling elements, e.g. UML components and UML constraints. The second extension (Fig. 4.8, right hand side) consists in adding an attribute criticality to the $NFP\_CommonType$ so that it can be used in VSL values.

These two small extensions impacts lots of modeling constructs since constraints and values are used in many different contexts. More precisely, the first extension adds a criticality level, through UML constraints, to application and platform components,

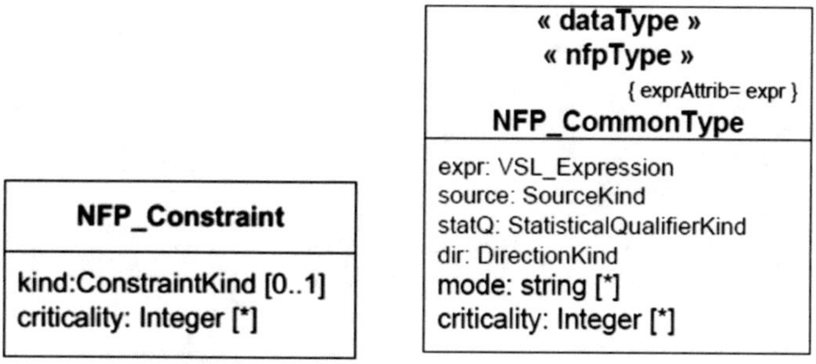

« dataType »
« nfpType »
{ exprAttrib= expr }
NFP_CommonType

expr: VSL_Expression
source: SourceKind
statQ: StatisticalQualifierKind
dir: DirectionKind
mode: string [*]
criticality: Integer [*]

NFP_Constraint

kind:ConstraintKind [0..1]
criticality: Integer [*]

Fig. 4.8 Extensions of MARTE for mixed-criticality

as well as performance requirements. The second extension adds a criticality level to value annotations, e.g. a WCET, and also to performance constraints captured. These modeling constructs provide solutions to cover different mixed-criticality modeling scenarios which have been identified and described in [16, 38].

Figure 4.9 shows an illustrative and novel MCS modeling scenario, where it is possible to add criticality levels along with the performance requirements of the system. Specifically, in Fig. 4.9, a criticality level 3 is assigned to a power performance requirement for the quadcopter system, i.e., the first kind of extension proposed.

Figure 4.10 completes the association of criticalities to other time-related performance requirements imposed to the quadcopter system. Specifically, the quadcopter model captures a number of deadline requirements on components of the quadcopter Platform-Independent Model (PIM). Those deadlines are captured as VSL value annotation through the attribute *relDl* of stereotype *RtSpecification*. The annotation of these performance requirements relies on our second proposed extension.

As mentioned, the single-source methodology covers other scenarios, e.g. modeling for mixed-criticality aware schedulability analysis, and for validation of architectural mapping according the criticality associated with application components and to the platform resources the former are mapped to (this is illustrated in Sect. 4.4.2).

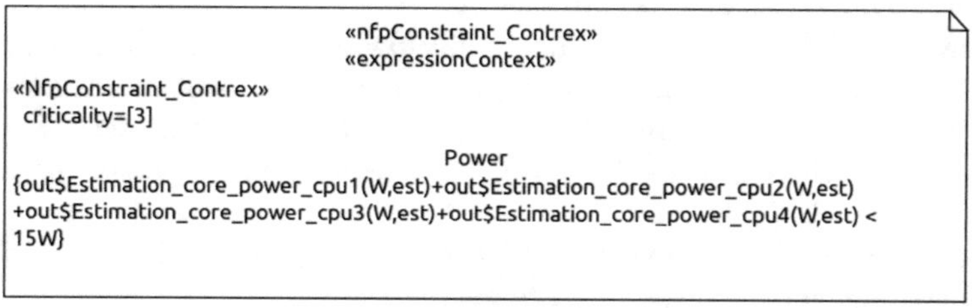

«nfpConstraint_Contrex»
«expressionContext»

«NfpConstraint_Contrex»
  criticality=[3]

Power
{out$Estimation_core_power_cpu1(W,est)+out$Estimation_core_power_cpu2(W,est)
+out$Estimation_core_power_cpu3(W,est)+out$Estimation_core_power_cpu4(W,est) <
15W}

Fig. 4.9 Criticality level assigned to a power-related requirement

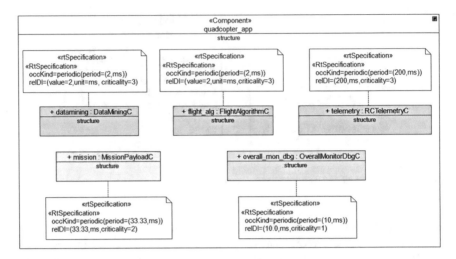

Fig. 4.10 Criticality associated with deadline requirements in the PIM

4.4 Tooling

4.4.1 State-of-the-Art

All the commercial UML editors support MARTE to diverse degrees. The simplest support consists in defining all the stereotypes. Others go further by usually allowing transformations of subsets into ad-hoc external tools. For instance, when using the *analysis* part of MARTE, one can performance schedulability analysis by transformations to RapidRMA or Cheddar (see http://www.omgmarte.org). Also another transformation from another subset of MARTE allows for performance evaluation [11].

Another possible set of analyses is available through the transformation [12] of a MARTE specification into AADL, which is a standard from the Society of Automotive Engineers and that gains interests for the design of safety-critical systems. There again, a specific subset of MARTE is retained.

TimeSquare [10] proposes a different family of tools. One side, when using clocks and clock constraints, TimeSquare attempts to build one behavior consistent with all the constraints. If this is possible, then it proposes one simulation model and animate the UML model. It can also use the constraints to detect possible inconsistencies and potential deadlocks [27].

Another approach [5] is to use a model-based approach to verify a given implementation through the analysis of execution traces.

A more detailed survey [6] gives an overview of other approaches based on UML, including MARTE. However, most solutions use ad-hoc transformations from a relatively small subset of UML and MARTE to ad-hoc external tools. In the following subsection, we promote integrated approaches with a single-source design for MARTE.

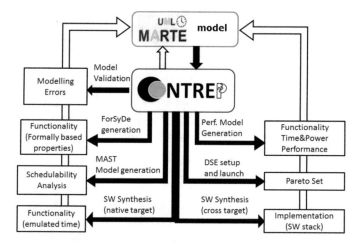

Fig. 4.11 The CONTREX Eclipse-Plugin provides a front-end and integrates tool infrastructure for single-source design from UML/MARTE

It is an attempt to bring consistencies among different views and integrate several tools within the same framework.

4.4.2 Single-Source Design from MARTE

The generic single-source design approach shown in Fig. 4.6 has been implemented by the University of Cantabria into the framework sketched in Fig. 4.11. In this framework, the CONTREP Eclipse Plugin (CONTREP) enables an unified front-end where the single-source model is developed, and where system-level design activities (boxes in Fig. 4.11) are applied from.

The architecture of the single design framework, revealing how and which profiles, libraries and tools are stacked is sketched in Fig. 4.12. Following, the tools integrated in the implemented framework and their role is explained under the perspective of the design tasks they enable. *Eclipse* MDT is at the base of the unifying front-end.

For the *modelling task*, *Papyrus* and the *MARTE profile* are at the base of the UML/MARTE modelling. CONTREP complements Eclipse/Papyrus with a profile with the MARTE extensions introduced in Sect. 4.3.2. CONTREP also provides a tool and a related configuration tab for the static validation of the model.

For enabling the different system-level design activities from the UML/MARTE model, CONTREP provides code generators, configuration and launching facilities. This performs an effective integration of state-of-the art and novel system-level design tools.

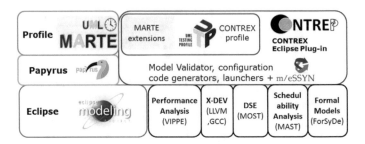

Fig. 4.12 Architecture of the single-source design environment

For *formally based functional validation*, CONTREP enables the generation of a ForSyDe model-SystemC model. By relying on ForSyDe-SystemC library [19, 28], CONTREP enables the generation of an executable model abiding the rules of the Synchronous Dataflow (SDF) Model-of-Computation (MoC) [20]. This ensures functional determinism by construction and makes analisable if the application is free from deadlock. CONTREP enables *schedulability analysis* as it integrates the *Marte2mast* code generator [14], able to produce a model that can be processed by the *MAST* [15] schedulabiltiy analysis tool.

CONTREP enables *software synthesis* for heterogeneous multi-core targets. For a given specific solution (once a platform and the architectural mapping is fixed), the platform dependent code is automatically generated, and the target binaries generated. In CONTREP, the code generators developed for eSSYN [37] have been integrated, extended and made available through its user menu.

The single-source design framework also enables the *automatic generation and the execution of an emulation model*. An emulation model enables the functional validation of the program, and the emulation of the timing specified on the application on top of the host machine. For instance, if the application states that a tasks will execute periodically, with period 1s, the produced emulated model will rely on the host RTOS services for emulating that timing behaviour. However, this type of model does not mimic the timing due to the mapping of the application to a given platform. In fact, the production of the emulation model considers the PIM, but neither the platform model nor the mapping information. The emulation model integrates platform independent functionality (referred by the PIM components of the model), and platform dependent functionality. The latter is generated through a specific code generator, the *m/eSSYN code generator*. This generator is invoked from CONTREP and produces all the code which implements the communication, concurrency and time services required by the semantics of the PIM components. In this sense, this platform dependent code *wraps* the platform independent code. Because of that, this code is also called *wrappers* code, and the code generator, *wrappers generator*.

CONTREP automates the generation of a fast executable performance model. The generation of the performance model relies on the VIPPE tool [39], which implements advanced techniques for fast simulation and performance assessment. The generation also relies on m/eSSYN, an extension of the eSSYN code generators, to support also a configurable simulation target. The m/eSSYN code generator allows to integrate DSE variables in the automatically produced code. This way, the performance model enables fast exploration (without trigerring new SW synthesis) of several design solutions considering also the variation of some aspect of the target dependent code, i.e. a task period. The final outcome is a more holistic exploration, in the sense that DSE parameters can be associated to platform attributes and also to the application. Simulation is convenient for getting accuracy and considering the dynamism of the application and of the input stimuli of the system environment, that is for an scenario-aware assessment of the system. VIPPE provides a rich set of metrics related to the time performance of the system. Thus the user is not only able to explore if the time requirements are fulfilled, but also facilitates the assessement of the main causes of eventual violations of those time requirements. Moreover, VIPPE provides energy and power consumption metrics. VIPPE relies on host-compiled simulation, and it is capable to parallelize the simulation to exploit multi-core host platforms. This is a main reason foor the suitability of VIPPE to be very suitable for design space exploration. Another one is that VIPPE supports the Multicube XML interface [35]. VIPPE is also capable to export power traces readable by tools able to perform dynamic temperature analysis, like *ThermalProfiler* [17].

CONTREP automates also the validation of a performance solution by automatically comparing the performance of a simulated solution versus the performance requirements captured in the UML/MARTE model. When the exploration is done in an interactive way, the comparison relies on two XML files. CONTREP code generators translate the performance requirements captured in UML/MARTE to an intermediate XML file. The run of the (VIPPE) executable performance model for a given combination of DSE parameter values (configuration or solution) produces an XML file reporting the performance metrics required to evaluate the performance requirements.

CONTREP also automates the generation of a complete simulation-based DSE infrastructure. As well as the performance model, files describing information like the design space, the performance constraints, the cost functions, and the exploration strategy, a basic input for the exploration tool coupled to the simulatable performance model is generated. The framework is flexible and allows the user to select the specific exploration tool to employ in the DSE (i.e. *MOST* [8] or *Multicube explorer*[40] so far), among the different exploration strategies available for the selected exploration tool, and among different report options, e.g., if intermediate results are preserved, of whether an HTML report is produced. Moreover, CONTREP allows to control the generation of a cost function which takes into account the criticality levels associated to the performance requirements. Thus the framework allows to give preference to solutions with larger safety margins in the more critical performance requirements, provided a minimum or equivalent performance on other metrics.

Both, for SW synthesis and for the production of an accurate VIPPE performance model, set ups of the cross-compiler tool-chains (X-DEV in Fig. 4.12) supporting the targeted platforms are required.

The final result is an joint user front-end to give access to triggering the different design activities from the same menu, shown in the capture of Fig. 4.13.

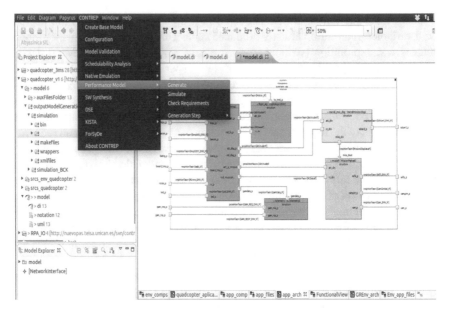

Fig. 4.13 Menu enabled by the CONTREX Eclipse Plug-in

4.5 Forecast About the Role MARTE May Have in Designing CPS

4.5.1 The End of Moore's Law

In its latest edition of 2015, the International Technology Roadmap for Semiconductors (ITRS) roadmap provided by the semiconductor companies changed its format from previous editions [18]. On the one hand, it confirms the end of Moore's Law as it has been performing until the first decade of the century. In fact, for several years now, the scaling has not been geometric as in the past. Although manufacturers still associate technology nodes with increasingly smaller dimensions, the size reduction is no longer the main cause of technological improvement. This improvement depends mostly on new structures (FinFET and FD-SoI transistors), new materials, etc. The roadmap proposes to achieve 10 nm technology in 2021. However, stagnation is observed from that date on. Only through technological improvements in processes with the same scale factor i.e. devices with the same size, will it be possible to manufacture integrated circuits with better performance. This is what is called equivalent scaling. A clear finale. An electronic technology that reaches maturity and is not going to be able to offer the exponential growth maintained to date. The end of Moore's Law will cause a radical change in the technological evolution as we have been enjoying it so far. On the one hand, the amortization of the billion investments to be made in each new line of semiconductor manufacturing can be spread on longer periods of time so that manufacturing costs can be significantly reduced. Today, advanced technologies are available only for those designs with high enough volumes of production. The aim is to ensure that the investment pays off in the short period of time in which that technology node is active before being replaced by more advanced technologies in new products. If this is no longer the case, semiconductor technology could become much more accessible to companies with smaller volumes. A similar evolution would follow design tools and libraries. On the other hand, for the designer, the pressure of time-to-market' is reduced, which will reduce design costs. Thus, it is quite possible that in the near future, the design of integrated circuits is accessible to any company in which the value added by silicon compensates the higher non-recurrent engineering cots. Between a product and its successor there will be only incremental improvements; no substantial improvements based on a significantly higher computing power, as it happens now. Innovation would come from the new applications and services based on new electronic products making use of application-specific integrated circuits fabricated with stable technologies. The price/performance evolution of electronic products will follow a similar trend to many other products. No longer fueled by the exponential improvement predicted by Moores Law. Some large companies are already preparing for this scenario. In some cases, this is an even desirable scenario as it could free up resources to invest in other technologies to offer new services to the marketplace.

4.5.2   The Rise of Connected Ubiquitous Smart Objects

Nowadays, we have just started to realize the enormous potential of an interconnected world of billions of smart devices providing new services to people. The end of Moore's Law commented above, might facilitate the proliferation of new electronic systems supporting these new services. Currently, most systems involve just a small number of computing resources, such as a data-center processing the voice from a smartphone and providing a voice-to-text service, or the distance sensors in a car connected to an Electronic Control Unit providing an automatic parking service to the driver. In these examples, specifying the complete service, deciding which functionality to execute in each node, and programming the corresponding application, although reasonably complex, are affordable tasks. However, services in a fully interconnected world will be composed of many SW components deployed on multiple devices of many kinds, from small sensing motes, embedded systems and smartphones to data-centers, and even, High-Performance Computing (HPC) facilities. All of them may be implemented with integrated systems containing heterogeneous processing elements (i.e. CPUs of different kinds, GPUs, DSPs and HW co-processors). And in all cases, the systems will need to satisfy functional and extra-functional critical constraints, including safety, security, power efficiency, performance, size, and cost. The global characteristics of the system as a whole will depend on the characteristics of their independent components, but also on the interactions with the physical environment and among them through the different communication networks. Therefore, the main innovation in the time to come shall be to jump from the design of cyber-physical systems (CPS) to cyber-physical systems of systems (CPSoS). These complex, heterogeneous, distributed systems require an interdisciplinary approach where the knowledge about the physical side of the systems is essential to arrive at solutions that are taken up in the real world. To integrate these diverse research and development communities is the most crucial aspect for a successful future development of CPSoS [9]. Current domain-specific methods are becoming obsolete; hence new predictive, engineering and programming methods and tools are required ensuring the satisfaction of the functional and extra-functional constraints imposed to the system while taking into account its interaction with the physical world. An additional important aspect to consider is the interaction between the system and the humans both as users or involved directly or indirectly in its operation (humans in the loop). The main reaction to the continuous evolution of computing platforms has been to decouple the application SW from the underlying HW. To achieve this goal many abstraction layers of middleware, communication protocols, operating systems, hypervisors and HW abstraction layers are being used. This approach is powerful enough for general purpose systems, for which extra-functional constraints such as execution times, energy efficiency, dependability, etc. are not strict. But the technological evolution towards CPSoS based on heterogeneous devices composed of CPUs of different kind, GPUs, DSPs, HW co-processors etc., added to the need to satisfy stricter non-functional properties makes this goal unrealizable. Although the software development on each of these platforms shares many commonalities, the current situation is that specific programming approaches, using different languages and tools are applied in each case. As a consequence, SW development for a supercomputer, for instance, becomes completely different to a smartphone or an embedded system. This was not a major problem when the supercomputer and the embedded system were functionally decoupled. Nevertheless, this is no longer valid in the fully connected world commented above. An initial functional partition followed by separate programming approaches leads to inefficient distribution of loads and communication traffic causing performance, energy, dependability, data movement, and cost overheads. Model-Driven Engineering (MDE) is a mature system engineering approach which is being successfully applied in many different domains. UML has proven its support to MDE in all these domains based on domain-specific profiles. Nevertheless, none of the currently available domain-specific approaches is applicable to develop SW for CPSoS or, the other way around, their applicability to different domains would require interoperability with other different languages, design methodologies and tools. Another important aspect is that in each domain the key concerns are also different. This is no longer valid. When dealing with a cross-domain application involving interacting SW components to be deployed and executed in different execution platforms, the functional and extra-functional properties (i.e. power consumption, real-time properties, performance, security, dependability, data movement) in one of the components may be affected by the rest of components. As a consequence, there is a need for a holistic modeling framework, across SW and HW layers, applications and domains. This modeling framework should be able to capture the complete high-abstraction model, integrating projects with different constraints (i.e. commercial or critical SW) and domains (i.e. from High Performance Computing, to embedded SW). The framework should integrate in a seamless and almost transparent way any abstraction layer (i.e. middleware, communication protocols, operating systems, hypervisors) required in a particular domain, on a particular HW platform. Currently fragmented but mature MDE techniques should be extended to the emergent, distributed, heterogeneous, CPSoS. Beyond the MDE underlying system engineering technology, the solution should allow an average skilled software developer to build an application for the advanced systems discussed previously. The goal is to avoid the system engineer a detailed knowledge of the characteristics of the underlying HW/SW platforms, letting him/her to focus on the Platform-Independent functionality of the complete system he/she wants to implement. The design framework should provide him/her with an accurate knowledge of the implications that the final implementation of the functionality on the concrete (distributed) platforms under a specific functional mapping will have in terms of extra-functional constraints. Beyond performance; energy consumption, safety, data traffic, security, adaptability, scalability, complexity management and cost-effectiveness have to be taken into account. This information about the complete system characteristics can be used in efficient optimization and design-space exploration for the complete system. As commented above, an essential aspect to be taken into account is the interaction of the system with humans all along the life-cycle, since the specification and design of the system until its deployment, field and obsolescence. UML has the potential to be the central modeling language in this new context. To achieve this goal, a consensus on a profile, powerful enough to capture all the relevant concepts required in CPSoS engineering while, at the same time, simple enough to find wide acceptance by the design community, is required. MARTE is a good starting point for two main reasons. Firstly, it captures most of the concepts required in system engineering on heterogeneous platforms under strict design constraints. Secondly, there is clear convergence among computing platforms and today, it is possible to find the same computing resources (i.e. CPUs, GPUs, and application-specific HW) in platforms apparently as different as an embedded system, a smartphone and a supercomputer. As the technology evolution will be incremental and no longer, exponential, this profile will be stable in time with only incremental improvements. If this is achieved, UML could become the system design language for the electronic century.

4.6 Conclusion

MARTE has the capability to describe several parts necessary to model CPS. Since MARTE is just a language, it does not come with a recommended or unique methodology. Here, we draw a map of what can be done with MARTE and what remains to be done. We advocate for a single-source modeling methodology where a central model is used to feed several analysis tools. This model relies on UML, MARTE, SysML along with very few extensions that we propose. The results are used to extend and refine the models following an iterative process.

To illustrate the process, we use the example of a quadcopter. We manage to put together several chapters (CoreElements, NFP, Time, Allocation, HLAM, VSL) of MARTE to propose a consistent usage in an unprecedented manner.

The question of whether UML is ill-founded is irrelevant, UML is there, it is widely used and will continue to be used. This is the only modeling language widely accepted by industry and that is expressive enough to cover so many aspects of complex heterogeneous systems. So, the good question, is rather to decide how to use it efficiently to gain an even bigger adoption and sound usage. We hope this chapter contributes to this objective.

Acknowledgements This chapter has been partially funded by the European FP7 611146 (CONTREX) project and by the Spanish TEC 2014-58036-C4-3-R (REBECCA) project. UC thanks the OFFIS team in CONTREX their support, documentation and material on their quadcopter implementation, which includes the quadcopter picture integrated in Fig. 4.4.

References

1. C. André, F. Mallet, R. de Simone, Modeling time(s), in *MODELS'07:10th International Conference on Model Driven Engineering Languages and Systems* Nashville, TN, USA, September 2007, Lecture Notes in Computer Science, vol. 4735 (Springer, ACM-IEEE), pp. 559–573
2. C. André, J. DeAntoni, F. Mallet, R. de Simone, *The Time Model of Logical Clocks Availablein the OMG MARTE Profile*, Chap. 7. (Springer Science+Business Media LLC, 2010), pp. 201–227, http://hal.inria.fr/inria-00495664
3. B. Bailey, G. Martin, A. Piziali, *ESL Design and Verification: A Prescription for Electronic System Level Methodology* (Morgan Kaufmann/Elsevier, San Francisco, 2007)
4. G. Berry, The Informatics of Time and Events. Collège de France, inaugural Lecture, March 2013
5. M. Bourdellès, S. Li, I. Quadri, E. Brosse, A. Sadovykh, E. Gaudin, F. Mallet, A. Goknil, D. George, J. Kreku, *Fostering Analysis from Industrial Embedded Systemis Modeling*, Chap. 11 (IGI-Global, Hershey, 2014), pp. 283–300
6. F. Boutekkouk, M. Benmohammed, S. Bilavarn, M. Auguin, UML2.0 profiles for embedded systems and systems on a chip (SOCS). J. Object Technol. **8**(1), 135–157 (2009), http://dx.doi.org/10.5381/jot.2009.8.1.a1
7. A. Burns, R. Davis, Mixed-criticality systems: a review, Technical report of Computer Science, University of York, 6th edn., August 2015
8. F. Castro, G. Palermo, C. Silvano, V. Zaccaria, MOST: multi-objective system tuner design space exploration for system architects, in *Proceedings of the Designing for Embedded Parallel Computing Platforms: Architectures, Design Tools, and Applications Workshop*, March 2011
9. CPSoS Working Group Members: Cyber-physical systems of systems: Research and innovation priorities book (2016), http://www.cpsos.eu/wp-content/uploads/2015/02/CPSoS-Provisional-Roadmap-Paper-for-public-consultation_web.pdf
10. J. Deantoni, F. Mallet, Timesquare: treat your models with logical time, in *TOOLS (50)*, vol. 7304, ed. by C.A. Furia, S. Nanz, Lecture Notes in Computer Science (Springer 2012), pp. 34–41

11. H. Espinoza, H. Dubois, S. Gérard, J.L.M. Pasaje, D.C. Petriu, C.M. Woodside, Annotating UML models with non-functional properties for quantitative analysis, in *Workshops of MoDELS 2005 Conference*. Lecture Notes in Computer Science, vol. 3844 (Springer, 2005), pp. 79–90, http://dx.doi.org/10.1007/11663430_9

12. M. Faugère, T. Bourbeau, R. de Simone, S. Gérard, MARTE: also an UML profile for modeling AADL applications, in *ICECCS* (2007), pp. 359–364

13. S. Friedenthal, A. Moore, R. Steiner, *A Practical Guide to SysML: The Systems Modeling Language* (MK/OMG, Burlington, 2014)

14. A. Garcia, J. Medina, MARTE2MAST, http://mast.unican.es/umlmast/marte2mast/

15. M. Gonzalez, J.J. Gutierrez, J.C. Palencia, J.M. Drake, Mast: modeling and analysis suite for real time applications, in *13th Euromicro Conference on Real-Time Systems* (2001), pp. 125–134

16. F. Herrera, J. Medina, E. Villar, Modelling hardware/software embedded systems with uml/marte: a single-source design approach, in *Handbook of Hardware/Software Codesign*, Chap. 5, ed. by S. Ha, J. Teich (Springer), pp. 125–159. The address of the publisher, 1 edn. (2 2017), printed version scheduled for Feb. 2017

17. Intel DOCEA: Thermal Profiler website, http://www.doceapower.com/index.php?option=com_content&view=article&id=237&Itemid=145

18. International roadmap for semiconductors. Technical report (2015), http://www.itrs2.net/

19. KTH Royal Institute of Technology: ForSyDe website (2016), https://forsyde.ict.kth.se/trac

20. E.A. Lee, D. Messerschmitt, Synchronous data flow (1987)

21. E.A. Lee, A.L. Sangiovanni-Vincentelli, A framework for comparing models of computation. IEEE Trans. Comput. Aided Des. Integr. Circ. Syst. **17**(12), 1217–1229 (1998)

22. E.A. Lee, Cyber physical systems: design challenges, in *11th IEEE International Symposium on Object-Oriented Real-Time Distributed Computing* (ISORC 2008) (IEEE Computer Society, May 2008), pp. 363–369, http://dx.doi.org/10.1109/ISORC.2008.25

23. E.A. Lee, S.A. Seshia, Introduction to Embedded Systems - A Cyber-Physical Systems Approach (LeeSeshia.org, 2014), ISBN 978-0-557-70857-4

24. M. Lemke, Mixed criticality systems, report from the workshop on mixed criticality systems, Technical report Information Society and Media Directorate-General, February 2012

25. F. Mallet, C. André, R. de Simone, CCSL: specifying clock constraints with UML/Marte. Innov. Syst. Softw. Eng. **4**(3), 309–314 (2008)

26. F. Mallet, Logical Time @ Work for the Modeling and Analysis of Embedded Systems (LAMBERT Academic Publishing, January 2011), ISBN: 978-3-8433-9388-1

27. F. Mallet, R. de Simone, Correctness issues on MARTE/CCSL constraints. Sci. Comput. Program. (2015). doi:10.1016/j.scico.2015.03.001

28. S.H.A. Niaki, I. Sander, An automated parallel simulation flow for heterogeneous embedded systems, in *Proceedings of the Conference on Design, Automation and Test in Europe* (DATE'2013), EDA Consortium, San Jose, CA, USA (2013), http://dl.acm.org/citation.cfm?id=2485288.2485297, pp. 27–30

29. OFFIS: CONTREX FP7 project website (2015), https://contrex.offis.de/home/

30. OMG: UML Profile for Schedulability, Performance, and Time Specification, v1.1. Object Management Group, January 2005. Accessed 02 Jan 2005

31. OMG: Systems Modeling Language (SysML) Specification, v1.1. Object Management Group, November 2008. Accessed 02 Nov 2008

32. OMG: UML Profile for MARTE, v1.1. Object Management Group, June 2011. Accessed 02 June 2011

33. OMG: UML Superstructure, v2.4.1. Object Management Group, May 2012. Accessed 07 May 2012

34. B. Selic, S. Gerard, *Modeling and Analysis of Real-Time and Embedded Systems with UML and MARTE* (Elsevier, Amsterdam, 2013)

35. C. Silvano, W. Fornaciari, G. Palermo, V. Zaccaria, F. Castro, M. Martinez, S. Bocchio, R. Zafalon, P. Avasare, G. Vanmeerbeeck et al., Multicube: multi-objective design space exploration of multi-core architectures. in *VLSI 2010 Annual Symposium* (Springer 2011), pp. 47–63

36. B.N, Taylor, A. Thompson, International System of Units, v1.1. National Institute of Standards and Technology, March 2008
37. University of Cantabria. TEISA Department GESE group: essyn website (2016), http://www.eSSYN.com
38. University of Cantabria. TEISA Department GESE group: UC single-source modelling and design website (2016), https://umlmarte.teisa.unican.es
39. University of Cantabria. TEISA Department. GESE group: Vippe website (2016), https://vippe.teisa.unican.es
40. V. Zaccaria, G. Palermo, F. Castro, C. Silvano, G. Mariani, Multicube explorer: an open source framework for design space exploration of chip multi-processors. in *Proceedings of the International Conference on Architecture of Computing Systems (ARCS)*, Febraury 2010, pp. 1–7

Chapter 5
Combined Model Checking and Testing Create Confidence—A Case on Commercial Automotive Operating System

Toshiaki Aoki, Makoto Satoh, Mitsuhiro Tani, Kenro Yatake and Tomoji Kishi

Abstract The safety and reliability of automotive systems are becoming a big concern in our daily life. Recently, a functional safety standard which specializes in automotive systems has been proposed by the ISO. In addition, electrical throttle systems have been inspected by NHTSA and NASA due to the unintended acceleration problems of Toyota's cars. In light of such recent circumstances, we are researching practical applications of formal methods to ensure the high quality of automotive operating systems. An operating system which we focus on is the one conforming to the OSEK/VDX standard. This chapter shows a case study where model checking is applied to a commercial automotive operating system. In this case study, the model checking is combined with testing in order to efficiently and effectively verify the operating system. As a result, we gained the confidence that the quality of the operating system is very high.

Keywords Model checking · Testing · Automotive systems · Operating systems

T. Aoki (✉) · K. Yatake
Japan Advanced Institute of Science and Technology, 1-1 Asahidai,
Nomi, Ishikawa 923-1292, Japan
e-mail: toshiaki@jaist.ac.jp

K. Yatake
e-mail: k-yatake@jaist.ac.jp

M. Satoh
Renesas System Design Co., Ltd., 3-1, Kinko-cho, Kanagawa-ku,
Yokohama, Kanagawa, Japan
e-mail: makoto.sato.jz@renesas.com

M. Tani
DENSO CORPORATION, 1-1, Showa-cho, Kariya, Aichi 448-8661, Japan
e-mail: tani@eeda.denso.co.jp

T. Kishi
Faculty of Science and Engineering, Waseda University, 3-4-1, Okubo,
Shinjuku, Tokyo 169-8555, Japan
e-mail: kishi@waseda.jp

© Springer Nature Singapore Pte Ltd. 2017
S. Nakajima et al. (eds.), *Cyber-Physical System Design from an Architecture Analysis Viewpoint*, DOI 10.1007/978-981-10-4436-6_5

5.1 Introduction

Recently, the safety and reliability of automotive systems are becoming a large concern in society. Although vehicles have been controlled by simple mechanics in the past, many of electronic parts are embedded in them at present according to the progress of electronic control technology and its performance. These electronic parts can actualize the complex control of the vehicles, and make it possible to provide high functionality to vehicles such as automatic speed controlling and emergency braking. The electronic control technology makes the vehicles more convenient and safer. Unfortunately, electronic parts also introduce the problems of the reliability and safety of the vehicles because the automotive systems become more complicated and their scale larger. In fact, highly electronized automotive systems have received much attention with respect to their reliability and safety. A functional safety standard which specializes in automotive systems has been proposed by the ISO [1]. Electronic throttle systems have been inspected by NHTSA and NASA because of the unintended acceleration problem of Toyota's cars in 2010 [2].

We are working on the verification of automotive operating systems to ensure the high quality of automotive operating systems. An operating system which we focus on is the one conforming to the OSEK/VDX [3] standard. OSEK/VDX is an organization which was established in 1993 and provides the industrial standards of ECU (Electronic Control Unit) architectures. OSEK/VDX deals with many kinds of components used in automotive systems and one of them is an operating system. Although AUTOSAR [4] takes over this activity, the OSEK/VDX standards is still used for automotive operating systems in practice. We use OS for the abbreviation of 'operating system' below.

Our purpose is to provide a high quality OS by applying formal methods which are recommended in the functional safety standards. OS has much impact on their safety evaluation because it is the base of automotive software which is embedded into automotive systems. JAIST and DENSO started a joint research project in 2006. DENSO develops automotive software using OSs which are provided by the other companies. We examined the feasibility of applications of formal methods at this point. Then, we decided to apply formal methods to a commercial OS whose target CPU is V850. Renesas Electronics Corporation (REL) which develops this OS and CPU joined this project in 2009. We call the OS 'REL OS' below. REL OS has been already released and used in a current series of cars at this time. It is needless to say that traditional methods have been applied to REL OS in order to check it then. Our aim is to achieve higher quality of the OS for next series of cars by applying formal methods.

This chapter shows a case study that model checking, which is one of formal methods, is applied to a commercial OS, that is, REL OS. REL OS is too complicated to convince us that it correctly performs for any application. We adopted exhaustive verification techniques to check REL OS. We have conducted exhaustive testing based on a design model which was exhaustively verified by model checking. As a result, we acquired the confidence that REL OS correctly performs for any application although no new bug was found since the model checking and testing were more exhaustive and reliable than the traditional methods. Such combined model checking and testing are appropriate to convince us of the correctness thanks to their exhaustive nature.

The rest of the chapter is organized as follows. We briefly introduce OSEK/VDX OSs in Sect. 5.2. Section 5.3 shows the overview of our approach to apply model checking and testing to the verification of REL OS. Section 5.4 discusses related works. Sections 5.5 and 5.6 explain the details and results of the verification. Section 5.7 discusses the approach and results. Section 5.8 concludes this chapter.

5.2 OSEK/VDX Operating Systems

OSEK/VDX OS, shortly, OSEK OS adopted a priority based scheduling of multiple tasks. Mixing preemptive tasks with non-preemptive tasks is allowed. It provides API functions such as ActivateTask, TerminateTask, and ChainTask for controlling the execution of tasks. ActivateTask activates a task, TerminateTask terminates a task, and ChainTask activates a task after terminating a task. A concept named resource exists to manage shared resources. The resource is obtained and released by API functions GetResource and ReleaseResource respectively. Mutual exclusion of tasks which have access to the shared resource can be realized by these API functions. They adopt a priority ceiling protocol [5] in order to avoid a priority inversion problem. Interrupt service routines are invoked when the interrupts occurred. We abbreviate interrupt service routines as ISR below. A priority is assigned to an ISR as well as a task. Synchronizing the tasks by events and alarms which invoke tasks based on time are also provided in addition to the API functions shown above.

The primary function of OSEK OS is to schedule tasks and ISRs. It is not too much to say that OSEK OS is almost a scheduler. The example of OSEK OS scheduling is shown in Fig. 5.1. The horizontal lines appeared in this figure represent time passage of tasks, a resource, or interrupts. The vertical direction shows their priorities. If one line is located upper to another line, the priority of the former is higher than that of the latter. The horizontal direction from left to right expresses time passing. OSEK OS allows us to mix preemptive tasks with non-preemptive tasks. TASK H(P) and TASK L(P) are preemptive tasks and TASK M(NP) is non-preemptive Task in Fig. 5.1. INT H and INT L represent interrupts. Interrupt service routines are invoked when the interrupts raised. We abbreviate interrupt service routines as ISR below. RES M represents a resource whose priority is between those of INT H and INT L. The execution order of the tasks and ISRs are as follows.

1. The task TASK L(P) is activated initially.
2. When the task TASK L(P) obtains the resource RES M, the priority of TASK L(P) is raised to that of RES M temporarily by the priority ceiling protocol.

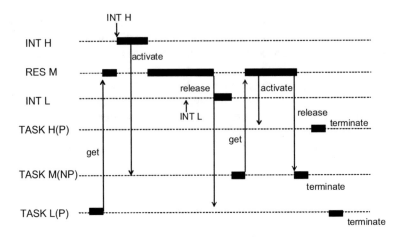

Fig. 5.1 Scheduling tasks in OSEK OS

3. When an interrupt INT H occurs, an ISR which corresponds to the interrupt is invoked after the preemption of TASK L(P). TASK M(NP) is activated inside the ISR.
4. When ISR is completed, TASK L(P) is resumed.
5. Even though an interrupt INT L occurs, while TASK L(P) is executing, an ISR which corresponds to that interrupt is not executed because the priority of INT L is lower than that of TASK L(P) at this point.
6. When TASK L(P) releases RES M, its priority becomes the original one. Then, TASK L(P) is preempted and an ISR of INT L is executed.
7. After the ISR of INT L is finished, the task TASK M(NP) is executed.
8. The priority of TASK M(NP) is temporarily raised to RES M by obtaining the resource RES M and it activated TASK H(P).
9. Even though the priority of TASK M(NP) becomes the original one by releasing RES M, it continues the execution since this task is non-preemptive.
10. After TASK M(NP) is terminated, TASK H(P) is executed.
11. After TASK H(P) is terminated, TASK L(P) is executed.

In the scheduler of OSEK OS, information needed for the scheduling is managed by data structures such as a queue and tables. The scheduler makes use of those data structures and calculates a task or ISR to be executed. Such calculation is very complicated since there are various configurations of priorities and preemptions, activation timings of tasks, and ISRs, and synchronization mechanisms. For example, in an OS conforming to OSEK/VDX, an activation of a task was ignored if an ISR activates the task whose priority was lower than another task preempted by the ISR. In this case, the scheduling of tasks becomes incorrect since a task TASK M(NP) is not executed at the eighth step even though ISR of INT H activates the task at the third step in Fig. 5.1. This incorrect scheduling is encountered not in all of configurations but in a specific one. It is very important to ensure that the computation is correct for any configuration. How the scheduling has to work is defined by OSEK/VDX standard specification. In our joint research project, we verified that the scheduler of REL OS surely conforms to the specification.

5.3   Approach

We show the overview of our approach in Fig. 5.2. Our approach is divided into two kinds of activities, design verification and testing. We have constructed a design model to clarify calculation carried out in REL OS. We confirmed that the calculation is correct by applying model checking to the design model. Incorrect scheduling as shown in Sect. 5.2 would be detected at this point. Then, we have conducted testing based on the design model in order to confirm that the implementation of REL OS conforms to the design model in which the correct computation was realized. The testing brings our confidence that the implementation is correct because it is actually operated. The activities associated with the design verification and the testing are surrounded by solid lines and dotted lines respectively.

5.3.1   Design Model

We constructed and verified a design model of REL OS to analyze its scheduling mechanisms. The design model was verified by a model checking tool Spin [6].

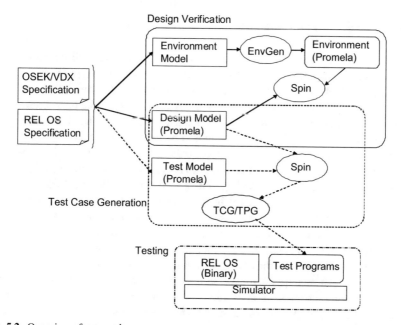

Fig. 5.2 Overview of approach

Spin checks properties represented as LTL formulas, assertions and so on against behavior represented as automata which are concurrently executed with channels to communicate with each other. Such behavior is described in a specification language named Promela. A Promela description consists of concurrent processes whose behavior is described as guarded commands in an operational way. It also provides various datatypes such as arrays and record types. Those datatypes allow us to straightforwardly describe the scheduling mechanisms of REL OS. In addition, the syntax of Promela is similar to C language which is familiar to engineers who develop REL OS. It is easy for the engineers to learn as well as communicate with researchers based on Promela descriptions. Thus, we constructed the design model in Promela.

5.3.2 Design Verification and Environment Modelling

5.3.2.1 Closing Open System by Environments

OSEK OS is an open system which performs when an API function is invoked by tasks or ISRs. It does not do anything if nothing is invoked. Similarly, the design model is not executable by itself. To check it by Spin, we need descriptions invoking API functions in addition to the design model. Such descriptions are usually called *environments* because it is outside of the design model.

5.3.2.2 Types of Environments

The environments manage invocations of functions to a target, callback from the target, inputs to the target and outputs from the target. There are two types of the environments. One is that the environments make completely non-deterministic invocations of functions and inputs to the target. The other is that they make non-deterministic invocations within specific execution contexts. The former is called *universal environments* [7]. Although the universal environments allow us to exhaustively check behavior of the target, many spurious errors will be reported. To avoid them, we need to constrain the behavior of the universal environment, for example, provide a filter of the behavior by LTL formulas [8]. Our objective of the design verification is to ensure the fact that a task selected by the scheduler is correct with respect to the specification. To represent this fact, we need to constrain the behavior of the environment so that it can make invocation sequences which lead to the selection of a specific task. However, it is very hard to describe LTL formulas representing those invocation sequences. In addition, the universal environment likely causes a state explosion problem. Thus, in our approach, we adopted the latter type of the environments, which make non-deterministic invocations within specific execution contexts. Those environments are represented as automata which define the contexts. The automata make it easier to describe the invocation sequences as the environments.

5.3.2.3 Facilitating Variations of Environments

In our approach, the environments are described as an environment model that we proposed in [9]. There are a number of variations of configurations such as the number of tasks, the number of resources, priorities of tasks, and ceiling priorities of resources for the environments. It is very hard to manually describe those environments one by one. Thus, we have proposed the environment model which represents variations of the configurations and allows us to automatically generate environments described in Promela. The variations are modeled in the class diagram of UML [10] and OCL (Object Constraint Language) [11]. The invocation sequences of API functions are modeled in the statechart diagram of UML with some extension. The expected results of the invocation of the API functions are also described in the statechart diagram with OCL. Then, we have developed a tool named *EnvGen* [9, 12] which automatically generates environments described in Promela from the environment model consisting of the class diagram and statechart diagram. The environments are generated within specific bounds of the variations of the configurations. The expected results are realized as assertions in the environments.

5.3.2.4 Constructing Environment Models

The possible configurations of the environments of REL OS are described in the environment model. Such configurations are identified from the specification of REL OS in addition to the OSEK/VDX standard specification. The invocation sequences and expected results of the API functions are also identified and described in the environment model similarly. In verifying the design model of REL OS, we generated environments from this environment model. The design model is coupled with each of the environments generated by EnvGen, then it is checked by Spin. That is, the design model is checked as many times as the number of the environments.

5.3.3 Testing

5.3.3.1 Confirming Conformance by Testing

We made much effort to ensure that the scheduling of tasks was correctly realized in the design model. What we had to do next was to ensure that the implementation of REL OS conforms to the design model. There are two approaches to ensure that an implementation conforms to a design model in general. One is that we generate a source code from the design model. The other is if the implementation conforms to the design model after manual implementation. We selected the second approach. A primary reason why we selected this approach was that REL OS has been already implemented. Another reason was that it was very hard to refine the design model so that a source code can be generated. REL OS is implemented in an assembly language of V850 to achieve high performance of executions. In addition, there are many mechanisms and optimizations, which are specific to V850. However, some such mechanisms do not appear in the design model since it focuses on the computation of the scheduling. Therefore, it would be very hard to refine the design model so that it can be isomorphic to the implementation of REL OS.

5.3.3.2   Regarding Design Model as Test Oracle

In our approach, we test the implementation of REL OS by test cases which are generated from its design model in order to check that the implementation conforms to the design model. We assigned much importance to the verification of the design model. We not only checked the design model by Spin but also reviewed the design model and environment model carefully. As the result of this effort, we assume that the design model is correct, that is, regard it as a test oracle. This assumption is reasonable since the design model must be relatively reliable in comparison to the other artifacts. Test cases are generated from the design model. Those test cases contain invocations of API functions in addition to expected results of them. Obtaining the expected results is possible because correct calculation is done (we assume this) in the design model.

5.3.3.3   Covering Implementation States

Conformance testing based on automata has been studied for a long time [13]. By these studies, it is well-known that strong assumptions are needed to decide that one automata conforms to another one. However, it is difficult for practical systems to discharge those assumptions. Therefore, we do not aim at this theoretical conformance but cover all the states, which appeared in the design model. In this approach, in order to cover states that we expect to test in the implementation, the design model needs to contain corresponding states. Thus, we constructed the design model so that if states of the implementation are different to each other, corresponding states of the design model can be also different to each other. This makes it possible to generate test cases, which reach expected states of the implementation. We use a model checking tool to obtain test cases which cover all the states of the design model.

5.3.3.4   Environments for Test Case Generation

We need environments in generating test cases as well as in the design verification since the design model does not compute anything if no API function is invoked. We call the environments to generate test cases *test models*. The test models are different from the environments of the design verification. The test models do not check the design model but only invoke API functions non-deterministically within some bounds.

5.3.3.5 Tools to Automate Testing

We have developed two tools named TCG (Test Case Generator) and TPG (Test Program Generator) for automatic testing based on the design model. TCG automatically generates test cases using Spin. Our approach is to generate test cases not by trap properties [14] but exhaustive search algorithm of states with a model checking tool. TCG generates test cases, which are reachable to all the states appearing in the combination of the design model and test models. Generated test cases consist of invocation sequences of the API functions and expected results. TPG transforms the test cases into programs to test REL OS. A program generated by TPG is compiled with REL OS and executed in a simulator and debugger of V850. TCG and TPG allow us to automatically perform testing of REL OS using the design model and test models as inputs.

5.4 Related Works

A word 'verification' is recognized as proving correctness with theorem provers or deductive techniques. The verification of OSs is challenging as demonstrated by the existing researches [15]. The verification of seL4 kernel is known as a recent notable success story [16, 17]. The word 'verification' is not limited to such deduction based approaches but used for model checking. Penix et al. [7, 18] verifies the time partitioning of DEOS. In this work, environments are obtained by filtering a universal environment with assumptions described in LTL. This approach is effective when the assumptions can be described simply, but shows weaknesses when describing precise behavior of environments because the assumptions described in LTL become complex. In our approach, we adopted different types of environments using automata to simply describe properties of scheduling. In addition, this work only verifies the design model despite that our approach deals with not only the verification of the design model but also the testing of the implementation.

There are several works on the verification of OSEK OS. Zhu et al. [19] verifies OSEK OS implemented in C language. The primary purpose of this work is to formally specify API functions of the OSEK OS. A part of such specifications is verified by VCC [20]. Huang et al. [21] manually constructs a model of CSP based on the source code of OSEK OS. Then, it is checked by a model checking tool PAT [22]. They do not take into account the conformance between the model and the source code. Choi [23, 24] verifies an open source OS named Trampoline [25], which is implemented in C language. In this work, the source code of Trampoline is analyzed by Spin. A model of Promela is manually constructed, then it is checked against properties obtained by safety analysis. This work does not take the conformance between the model and the source code into account as well. In comparison to those works, the originalities of our work can be summarized as follows. Firstly, our approach covers both of design and testing phases in developments although the other works focus on a single activity or phase of the developments. We combine the verification of the design model with testing of the implementation seamlessly. Secondly, our target is implemented in the assembly language of V850. Thus, we cannot take existing techniques which are specific to C language like [19, 23, 24].

In our previous researches, we proposed a tool to automatically generate environments [9] and it has been applied to a design model of OSEK OS [26]. We adopt the tool for verifying a design model of REL OS. The design model described in [9, 26] is different from that of REL OS. We have proposed an approach to automatically generate test cases from the design model [27]. In the approach, test scenarios to generate the test cases were described in Z notation [28]. We do not describe the test scenarios in this chapter. Instead of the test scenarios, we describe a test model which non-deterministically invokes API functions of OSEK OS to exhaustively generate test cases since our purpose of the verification is to obtain the confidence thanks to the exhaustive nature. In addition, we made trade-offs and decisions for obtaining the confidence throughout the design verification and testing. In this chapter, we show a practically integrated approach to obtain the confidence and experiences that we gained in the case study.

5.5   Design Model and Verification

5.5.1   Construction of Design Model

We constructed the design model of REL OS in Promela. As mentioned in Sect. 5.2, we focus on its scheduler. The scheduler of REL OS has data structures consisting of a ready queue, tables, and flags. The ready queue records activation orders of tasks for each of the priorities. A task to be executed is determined based on the ready queue. It is obtained by searching the highest and firstly activated task recorded in the ready queue. The tables record information of tasks such as the current states and priority of tasks. The flags record conditions needed for the scheduling. High performance of the scheduling is required for the OS since it controls machinery of automobiles. On the other hand, searching the ready queue and switching tasks are not efficient. To achieve high performance, the flags are used for identifying whether costly operations are needed or not. Such data structures can be straightforwardly described in Promela.

Figure 5.3 shows a part of the design model described in Promela. In the implementation of REL OS, the ready queue is realized as a specific memory area of ECU. Operations to enqueue/dequeue a task to/from the ready queue is implemented by instructions of V850 which calculate addresses to update the memory area. We did not model the ready queue based on such memory area and instructions. We modeled the ready queue as an array instead of the address calculation. In Fig. 5.3, the ready queue is represented by an array named 'ready' in the design model. Operations to enqueue and dequeue are described as inline macros which update the array.

Datatypes

```
#define N_PRIO_TASK 72 /* maximal tasks in a queue */
#define N_TASK 4 /* maximal tasks */
#define OS_ACT_MAX 2 /* maximal multiple activations */
#define TID byte /* task identifier as byte */
#define PRI byte /* priority as byte */
...

#define queue(x,y) ready[((x) * N_TASK * OS_ACT_MAX) + (y)]
TID ready[N_PRIO_TASK]; /* ready queue */

...
#define NOTEXIST      0
#define SUSPENDED      1
#define READY      2
#define RUN      3
#define WAITING      4
...

typedef TCB{
PRI tpriority; /* priority */
byte tstat; /* task state */
byte actcnt /* activation counter */
....
}
TCB tsk_state[N_TASK];
....
TID turn = EMPTY; /* context */
...
#define E_OK                      0
#define E_OS_ACCESS   1
#define E_OS_CALLEVEL 2
#define E_OS_ID          3
byte ercd; /* error code */
...
```

Basic operations on datatypes

```
inline enq(pr,id){
  enqueue 'id' in a queue of 'pr'
}
inline deq(pr,id)
  dequeue 'id' from a queue of 'pr'
}
.....
```

API functions

```
inline ActivateTask(t){
  error check;
  get an array index idx corresponding to t;
  if
  :: tsk_state[idx].actcnt < OS_ACT_MAX ->
    tsk_state[idx].actcnt++
    if
    :: tsk_state[idx].tstat == SUSPENDED ->
      enq(tsk_state[ret_ix].tpriority, id);
      tsk_state[idx].tsat = READY;
      ercd = E_OK;
      ....
    :: tsk_state[idx].stat == READY ->
      ....
    fi
  fi
}
inline TerminateTask(t){...}
inline ChainTask(t1,t2){...}
....
```

Fig. 5.3 Design model

TCB (Task Control Blocks) which hold information of tasks is represented by an array named 'tsk_state'.

If an API function is invoked, a task to be executed is determined after values of the datatypes are updated. For example, if ActivateTask(t) where t is a task identifier is invoked, t is enqueued into the corresponding position of the ready queue as well as a current state of a task t which is recorded in the tables is updated to a ready state. Then, a search of the ready queue and switching a task are performed if needed. Such operations are realized as inline macros of Promela in the design model as shown in Fig. 5.3. Interrupt handling mechanisms are described in the design model in addition to operations of API functions. The interrupts affect the scheduling of tasks although they are processed by hardware in actuality.

The design model has been constructed by the JAIST team in our joint project. The initial design model has been constructed based on the OSEK/VDX standard. However, its behavior was not the same to REL OS due to some misunderstandings and the ambiguity of the specification. It was reviewed by engineers of DENSO and REL to make its behavior equivalent to that of REL OS. We held regular meetings to consider review results once a month. The improvements of the design model were also done by the JAIST side according to the review results. It took around 6 months to improve it and we finally obtained the design model of REL OS.

5.5.2 Environment Modelling

As mentioned in Sect. 5.3, in our approach, the environments are described as an environment model that we have proposed. We describe our decisions and tradeoffs that we made with brief introduction of the environment model here. For more details, please refer to our earlier works [9, 12, 26]. Figure 5.4 shows an example of the environment model. The left-hand side of Fig. 5.4 models variations of configurations in the class diagram of UML. The set of tasks and resources are denoted as classes named *Task* and *Resource* respectively. The multiplicities are assigned to the classes since there can be multiple tasks in a configuration. M and N represent the maximal number of tasks and resources respectively. There are relations between them. For example, there is a relation that a task uses resources. Such relations are represented as an association between the classes *Task* and *Resource*. In addition, there is a constraint that if a task uses a resource, the ceiling priority of the resource has to be higher than or equal to the priority of the task. This constraint is described in OCL and attached to the class *Task*. In this way, we define the variations of configurations in the environment model.

Invocation sequences of API functions and expected results are described in a statechart diagram of UML and OCL as shown in the right-hand side of Fig. 5.4. This statechart diagram represents the collective behavior of tasks instantiated from the class *Task*. We introduce some extensions to describe such behavior. For example, `|Run->Rdy:GetRun()` is called a *synchronous transition* which triggers simultaneous transitions of multiple objects. This transition is needed for expressing the fact that state changes of a task affect that of another task.

The environment generator EnvGen generates all possible environments of the environment model. Environment generation is conducted in three steps. Firstly, all possible object graphs are generated within the bounds of the class model. In [9], we performed this generation by an elementary algorithm that enumerated all of the object graphs in alphabetical order. However, the performance of this algorithm is poor. Currently, we have updated the generator to use a satisfiability modulo theories (SMT) solver for enumerating all the object graphs [12]. In this version of EnvGen, the performance was improved very much. Secondly, for each object graph, we generate a labeled transition system (LTS) by composing the statechart diagrams of all objects in the object graph. Finally, we generate the environments by translating each LTS into a Promela file.

5.5.3 Approximation of Environments

The environment model represents the set of invocation sequences of API functions and their expected results. They are, in a sense, analogous to test cases and their expected results. We want to examine the design model ultimately for all the invocation sequences. However, if we construct the environment model which deals with

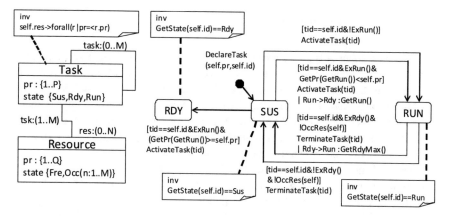

Fig. 5.4 Environment model

those invocation sequences, its complexity becomes similar to that of the design model. For example, if we describe expected results in the case where multiple tasks whose priorities are the same are activated, we need a queue like the ready queue of the design model in the environment model because activation orders of tasks have to be recorded in the environment model as well. Creating such an environment model makes little sense because its reliability becomes as uncertain as the design model. The reliability of the environment model should be higher than the design model from the viewpoint of practicality.

There are two approaches, over approximation and under approximation, to solve this problem. In the former, we construct an environment model so that it can contain more invocation sequences and stronger expected results than exact ones. For example, if there are multiple tasks whose priorities are the same and in the ready states, an expected result is that one of those tasks should be executed. In this case, we do not need a queue to describe expected results and make the environment model simpler. However, such expected results make little sense because they are too strong. For example, checking the fact shown above can not detect errors in the order to become running. Thus, it does not contribute to the correctness of the scheduling very much. In addition, many false positive counter examples would be reported because the environment model contains many of non-deterministic invocations of API functions. Hence, we took the latter approach which is under approximation. In this approach, we construct an environment model so that it can contain less invocation sequences and weaker expected results than exact ones. This under approximation is achieved by case splittings of environment models. Thus, we constructed the environment models separately such that each model was as simple as possible. This separation allows us to provide high reliability to the environment models. In addition, the separation of models can reduce the risk of state explosion. If we check all aspects of the design model at once, state explosion can easily occur. We can check each of them within a relatively small state space by separating the environment models.

5.5.4 Verification Result

We construct the environment models separately from each other. Such separation introduces a problem of coverage. This approach does not cover all the invocation sequences of API functions because it is based on the under approximation. On the other hand, as discussed so far, neither fully non-deterministic nor over approximated invocations of API function can be accepted in our approach. Thus, we decided to take the under approximation approach and carefully separate the environment models so that they can cover our concerns in the design model.

Figure 5.5 shows environment models that we have constructed. The environment models are divided into six groups based on which the functionalities of OSEK OS are being checked. Each group is further divided into two cases based on the equality of task priorities. For example, environment models No. 1 (TaskDiff) and No. 2 (TaskEq) check task management functions. They represent the cases with different priorities and the same priority respectively.

Figure 5.6 shows the results of environment generation and model checking. We used a computer whose specification is Intel Core2Duo CPU 2.4 GHz with 4 Gbyte memory. The environment generation results show the number of environments generated from each environment model, the time taken for generation, the average length of the Promela descriptions, and the average number of states and transitions contained in each environment. The model checking results show the time taken for checking all of the environments. We limited the number of tasks, resources, and ISRs to a maximum of 3. With these ranges, we were able to generate a total of 789 environments in about 100 s, which is quite efficient since only about 0.1 s was needed to generate each environment. This result demonstrates the effectiveness of using the SMT solver. In our previous work, the environments were generated by simply enumerating instances which meet constraints appeared in the environment models. Although the generation of the environments were impossible for three tasks and three resources, it was completed thanks to the SMT solver. For model checking, we were able to check the design model using all of the environments without state

No.	Name	Purpose	Condition
1	TaskDiff	termination and activation of tasks	different priorities
2	TaskEq	termination and activation of tasks	same priorities
3	CtDiff	ChainTask	different priorities
4	CtEq	ChainTask	same priorities
5	MultDiff	multiple activation of tasks	different priorities
6	MultEq	multiple activation of tasks	same priorities
7	ResDiff	get and release of resources	different priorities
8	ResEq	get and release of resources	same priorities
9	EvDiff	events	different priorities
10	EvEq	events	same priorities
11	IsrDiff	ISR	different priorities
12	IsrEq	ISR	same priorities

Fig. 5.5 Separated environment models

No.	name	generated environments					model checking
		num.	time(s)	lines	states	trans.	time(s)
1	TaskDiff	26	0.6	153	4	9	115.6
2	TaskEq	9	0.3	245	9	16	44.2
3	CtDiff	26	0.8	168	4	13	116.3
4	CtEq	9	0.3	273	9	23	45.9
5	MultDiff	26	0.9	199	8	19	119.7
6	MultEq	9	1.1	576	50	98	68.3
7	ResDiff	341	62.7	892	44	112	3828.5
8	ResEq	63	10.9	926	44	112	834.4
9	EvDiff	26	1.4	336	15	47	143.3
10	EvEq	9	1.6	1284	78	261	179.7
11	IsrDiff	182	13.9	502	17	49	1245.0
12	IsrEq	63	5.9	789	34	99	617.0

Fig. 5.6 Environment generation and model checking results

explosion occurring due to the separation of the environment models. The entire model checking took 122 min such that about 10 s per environment was required on average. Most of this time was used for compilation, which grows exponentially with the length of the Promela descriptions.

We conducted model checking several times while we were constructing the design model. The results shown in Fig. 5.6 are final ones. The design model was constructed and checked in this way by the JAIST side. We found many errors such as incorrect conditions and incorrect updates of data by model checking during its construction and verification. However, they were not the errors of REL OS but of the design model itself. That is, we described the incorrect conditions and updates which do not appear in REL OS in constructing the design model. Finally, no error was reported by model checking. Nonetheless, we gained confidence that the design model was highly reliable. We encountered many errors detected by our approach even though they were not the errors of REL OS. It made us believe that it was powerful enough to find errors.

5.6   Testing Based on Design Model

5.6.1   Generation of Test Cases and Programs

We made much effort to verify the design model by review and model checking. Then, we assume that it is correct and generate test cases with their expected results from the design model. We have proposed a method to automatically generate test cases from Promela descriptions by Spin [27]. In this method, search paths during model checking are recorded as search logs. Spin has an option to generate debug information such as up and down operations of depth-first search of model checking algorithm. In addition, we can print out information about the status of the design model such as invoked API functions and the current states of tasks during model checking thanks to the embedded C function of Promela. Such debug information and status make it possible to restructure a search tree of model checking in which expected results are contained. We obtain test cases, which are reachable to all the states of the design model and test models by scanning this search tree. We have developed a tool named TCG for testing REL OS according to this method. TCG inputs the design model and a test model. Then, it outputs invocation sequences of API functions and their expected results consisting of current states and priorities of tasks.

The test cases generated by TCG are not programs but the invocation sequences and expected results. Thus, we developed a tool named TPG which translates them into programs to be compiled with REL OS. The programs generated by TPG are regarded as applications executed on REL OS. The programs invoke API functions according to the test cases. In addition, they have statements to check whether status of REL OS is the same to those expected results. Such a check is realized by debugger of a V850 development environment. The results of the check is stored in a log file of testing.

TCG and TPG allow us to automate testing of REL OS based on the design model. If we give the design model with a test model, testing is automatically performed and then its results are recorded in the log file. In our project, TCG and TPG were developed by the JAIST side and REL side respectively.

5.6.2 Test Models

We need environments in generating test cases as well as the design verification since the design model does not do any computation if no API function is invoked. The environments to generate the test cases are called test models. Figure 5.7 shows a general form of the test models. A test model invokes API functions of the design model non-deterministically. The reason why preconditions are described is to prevent infeasible test cases from being generated. For example, invoking TerminateTask() by a task whose state is not running is infeasible in an actual execution. Such invocations are excluded by the preconditions. Execution context of the design model is needed to describe preconditions. For example, the current states of tasks are needed to describe the precondition of the invocation of TerminateTask(). This execution context can

Fig. 5.7 Test model

```
do
:: precondition$_1$ -> API function$_1$
:: precondition$_2$ -> API function$_2$
...
:: pre-condition$_n$ -> API function$_n$
od
```

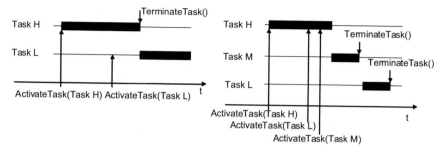

Fig. 5.8 Task switches

be obtained by referring to states of the design model. This is reasonable because we assume that the design model is correct for conducting the testing. Checking the design model is not an objective in testing. Reference to the states of the design model make it easier to describe test models.

Configurations have to be bounded in the test models. We have to determine the number of tasks, resources, ISRs, their priority assignments and events for the generation of test cases. We investigated behavior of task switches realized by REL OS since we focus on its scheduling. We describe each of possible variations of task switches as a use case. Then, we consider that what configurations cover those variations. For example, we show two of use cases in Fig. 5.8. Two tasks which have different priorities are sufficient to ensure the fact that a task whose priority is the highest among tasks whose states are ready. This case is shown in the left-hand side of Fig. 5.8. To ensure the fact that a task to be executed does not depend on activation orders, we need three tasks which have different priorities. This case is shown in the right-hand side of Fig. 5.8. In this way, we identified the following configurations which cover the variations of task switches. The numbers of tasks, resources, ISRs and events are 3, 2, 1, and 1 respectively.

5.6.3   Test Cases and Test Programs

Figure 5.9 shows a test case and test program generated by TCG and TPG respectively. The test case represents an invocation order of API functions as follows.

1. A task named task1 invokes ActivateTask(task2).
2. A task named task2 invokes ActivateTask(task3).
3. An interrupt whose number is 1 occurs.
4. An interrupt service routine isr1 invokes SetEvent(task1, Event1).
5. An interrupt which occurs in 3 is reset.
6. ⋯

The expected current states such as ready queue and TCB exist in the test case, however; they are omitted here for the sake of simplicity. Timings that interrupts occur are also described. SetINTR(1) and ResetINTR(1) represent that the interrupt which triggers an interrupt service routine named isr1 is set and reset respectively.

The test program realizes the invocation order described in the test case as well as checks the expected current states. Although the test program consists of tasks, the test case is an invocation sequence of API calls. Thus, we need to transform the test case to the tasks which cause the invocation sequence. The test case contains information which makes it possible to assign invocations of API functions to tasks. The invocation order of assigned API functions has to be controlled inside of each of the tasks. The variables exccnt1 to exccnt4 control the invocation order so that it can follow the one represented in the test case. The expected current states and execution order of API functions are checked using debug functions of the simulator. The timings that the interrupts occur are controlled by a library of the simulator. Code fragments to check them and control the interrupts exist in the test program, however; those are omitted in Fig. 5.9.

Test Case

```
task1:ActivateTask(task2) task2:ActivateTask(task3) SetINTR(1)
isr1:SetEvent(task1,Event1) ResetINTR(1) task3:TerminateTask()
task2:TerminateTask() task1:ActivateTask(task3) SetINTR(1)
isr1:SetEvent(task3,Event1) isr1:ActivateTask(task2) ResetINTR(1)
task3:TerminateTask() task2:TerminateTask() task1:ChainTask(task2)
task2:ActivateTask(task1) task2:TerminateTask() task1:ActivateTask(task3) ...
```

Test Program

```
ISR(isr1){
 if(exccnt1 == 1){
  ercd = SetEvent(task1,Event1);
  return;
 } if(exccnt1 == 2){
  ercd = SetEvent(task3,Event1);
  ercd = ActivateTask(task2);
  return;
 }
 if(exccnt1 == 3){
 /* if branches continue */
 }
}

TASK(task1){
  ercd = ActivateTask(task2);
  ercd = ActivateTask(task3);
  ercd = ChainTask(task2);
  return;
 }
 if(exccnt2 == 2){
 exccnt2++;
  ercd = ActivateTask(task3);
  ercd = ActivateTask(task3);
  ercd = ChainTask(task2);
  return;
 }
 if(exccnt2 == 3){
 /* if branches continue */
 }
}
```

```
TASK(task2){
 if(exccnt3 == 1){
  exccnt3++;
  ercd = ActivateTask(task3);
  ercd = TerminateTask();
  return;
 /* if branches continue */
 }
 if(exccnt3 == 2){
  ercd = TerminateTask();
  return;
 }
 if(exccnt3 == 3){
 /* if branches continue */
 }
}

TASK(task3){
 if(exccnt4 == 1){
  exccnt4++;
  ercd = TerminateTask();
  return;
 }
 if(exccnt4 == 2){
  exccnt4++;
  ercd = SetIntr(1);
  ercd = TerminateTask();
  return;
 }
 if(exccnt4 == 3){
 /* if branches continue */
 }
}
```

Fig. 5.9 Test case and test program

5.6.4 Test Results

Figure 5.10 shows results of generating test cases and programs by TCG and TPG. Three tasks and two resources are named TaskA, TaskB, TaskC, ResourceA, and ResourceB respectively. An ISR and event are omitted in this figure. Priority assignments of the tasks and resources are described in their rows. For example, upper-left of Fig. 5.10 represents that the priorities of TaskA, TaskB, and TaskC are 1, 2, and 3 respectively. Furthermore, there exist six variations of ceiling priorities of the resources. Regarding to these values, greater values mean higher priorities. Note that variations of priority assignments are reduced by considering their symmetry. The row of '#test cases' represents the number of test cases generated by TCG. The rows of 'pan exe. time', 'TCG exe. time', and 'TPG exe. time' represent amounts of time which are taken to search reachable states by Spin, generate test cases by TCG and generate test programs by TPG respectively where time units are seconds. We generated the test cases and test programs by a computer whose specification is Intel(R)Core2Duo CPU 3.00 GHz with 1 Gbyte memory. The total number of test programs generated by these configurations is 742,748.

Each of the generated test programs was compiled with REL OS and executed on the simulator of V850. These test programs can be executed independently. Thus, it is possible to perform testing in parallel in principle. However, we used a debugger to check REL OS and the number of its licenses that we can use is limited to three. In addition, some of the licenses are often occupied by engineers of REL and we need to exclusively use them. Therefore, we executed the test programs in a single computer in the daytime of weekdays, and in parallel on weekends and in the midnight of weekdays. As a result, we took around 3 months to complete checking all the test cases and no failure of test cases was found in the testing.

	Priorities						Priorities					
TaskA	1						1					
TaskB	2						1					
TaskC	3						1					
ResouceA	1	1	2	3	2	1	1	1	2	3	2	1
ResouceB	2	3	3	3	2	1	2	3	3	3	2	1
#Test cases	12483	15077	26373	37127	25035	8495	26489	26489	66361	66361	66361	13301
pan exe. time	12.9	16.8	26.0	38.7	26.7	10.7	27.6	30.6	66.3	66.9	68.3	14.6
TCG exe. time	19.0	24.9	67.9	73.1	58.5	12.7	81.8	93.7	289.2	287.2	282.6	27.2
TPG exe. time	176.8	174.4	508.5	522.8	433.9	95.0	548.4	626.9	2267.4	2694.1	2692.2	232.5
	Priorities						Priorities					
Task A	1						1					
TaskB	1						2					
TaskC	2						2					
ResouceA	1	1	2	3	2	1	1	1	2	3	2	1
ResouceB	2	3	3	3	2	1	2	3	3	3	2	1
#Test cases	17151	22427	44723	60707	39457	10331	13011	20707	33117	56281	25459	9425
pan exe. time	19.0	22.7	45.0	60.5	40.2	11.5	13.7	21.9	33.4	55.6	25.5	9.9
TCG exe. time	35.6	44.7	117.6	179.7	99.4	16.8	21.8	38.8	78.3	161.8	55.1	14.1
TPG exe. time	290.7	320.2	799.8	1353.5	694.4	138.9	175.2	317.9	546.8	1486.5	442.8	125.3

Fig. 5.10 Test results

We could not measure exact time taken to check all of them because testing was parallelize in an ad-hoc way. As a representative, we measured time taken to execute a part of the test programs instead. For example, it took 165.75 h to check 26,489 test programs where time to compile and execute them are 42.5 and 127.25 h respectively. It took 265 h to check 44,723 test programs where time to compile and execute them are 80.25 and 184.75 h. The test programs were complied and executed by a computer whose specification is Pentium4 3.2 GHz CPU with 1 Gbyte memory. We can estimate the whole of time to be taken to complete the execution of the test programs based on these data as around 4,535 h, that is, 189 days. We can say from this estimation that our parallelization (even though it is ad-hoc) contributed to the reduction of time taken for the testing because around 3 months were taken to complete it in fact.

5.7 Discussion

5.7.1 Practical Applications of Model Checking

An important technical achievement of this study is that we succeeded in seamlessly connecting two verification activities; verification of a design model with model checking and testing of a product. A point to achieve this is to regard a design model as a test oracle after making much effort to ensure its correctness with respect to the specification. We rely on the design model when we generate test cases and programs. In this approach, construction and verification of the design model are directly associated with testing of products which are recognized as an important activity in industries. That fact makes it easier to motivate engineers to construct a formal design model and use model checking. We could concentrate on constructing and verifying the design model by showing a way to effectively use them in developments.

It is ideal to apply formal methods to every phase of developments in principle; however, it is often not feasible in practice due to their high cost. Therefore, it is important to find an activity to be concentrated on and apply formal methods in the activity. In addition, making the best use of an artifact obtained by the application of formal methods in other activities is also important. Such concentration and effective use of the artifact make the cost to apply formal methods reasonable in practice. It is needless to say that this is a trade-off between theory and practice. The degree of correctness will be decreased in a sense, by restricting activities to apply formal methods. In our approach, we concentrate on the construction and verification of a design model in addition to adopting testing to check a product. As a result, we succeeded in keeping the cost reasonable so that model checking can be applied to a commercial product. The reason why we concentrated on the design model is that it is easy to characterize behavior of OSs like OSEK OS in an imperative specification language which is used to the design model. Actually, a ready queue appeared in the standard specification of OSEK/VDX to explain the behavior of the OSs.

Such precise behavior of the OS is taken into account from an early stage of developments. Thus, constructing a formal design model is relatively natural in development.

5.7.2 Verification Results

Our project consists of two researchers of JAIST and several engineers of REL and DENSO. The construction and verification of the design model were conducted by the JAIST side in which both of two researchers are involved. The review of the design model and verification results was conducted by all of members of this project. We took around 6 months to construct the design model. Although we did not measure the exact period to obtain the design model, it must be actually much less than 6 months since they had not only this project but also the other works. On the other hand, REL is developing this kind of OSs including REL OS more than ten years. In addition, REL OS was extensively verified by REL and DENSO before it was assembled in the current series of cars. Actually, they discovered and fixed bugs many times at that point. REL OS was already sufficiently checked when we started our project.

It was still surprising that no bug of REL OS was discovered in our verification because testing was conducted by a huge number of the test cases. Furthermore, we were surprised at the fact that the quality of the design model is as same as REL OS even though the development period of the design is shorter than that of REL OS. Remind that the design model was constructed by the researchers of the JAIST side who have little experience than the engineers of REL and DENSO with respect to OS developments. We initially, before the start of the testing, expected that if testing would fail, that should be due to bugs of the design model or misunderstanding of behavior of REL OS. However, all the test cases have been passed at a time. There was no backtrack to improve the design model once we started the testing. We succeeded in making the design model whose quality is similar to REL OS in 6 months. We can say from this fact that model checking effectively works for ensuring the quality of the design model. Although an objective of the testing is not to check the design model but the implementation, effectiveness to apply model checking to the design model has been proved consequently.

Our approach is based on exhaustive search methods. We encounter state explosion problem as far as we use the exhaustive search methods. Thus, we introduced techniques to prevent the state explosion problem as follows. Firstly, we bounded variations of configurations such as the numbers of tasks and resources in the verification of the design model. Secondly, we separated the possible behavior of environments into twelve cases. Finally, we bounded variations of configurations in the testing. We could not guarantee to cover the whole behavior of the design model and implementation due to the techniques. In addition, the verification of the design model relies on the environment models. Even though the environment models are simpler than the design model, they might still be incorrect. Our verification depends on the validity of such bounds and environment models.

To convince that the environment models and bounds are valid, we made effort to conduct manual inspection of the environment models and variations of configurations. There are techniques to make the validity more convincing. For example, we use theorem proving to verify the design model for unbounded variations. We construct a formal specification of OSEK OS, then verify the design model and environment model against it to make sure that they meet the specification. On the other hand, we have to pay additional cost if we adopt those techniques furthermore. We need to carefully decide what techniques we should choose from not only the theoretical but also practical point of view. The combination of techniques including model checking and testing as shown in this chapter is an approach whose cost is acceptable in the field of automotive systems.

5.7.3 Meaning of Testing

As mentioned so far, REL OS was already sufficiently checked by REL and DENSO. This check was conducted based on ordinary testing and review methods. Some bugs were found with respect to the scheduling of REL OS then. We confirmed that those bugs could be also detected by testing with our approach. As all the test cases have been passed, we can say that no such bug exists within behavior represented by them.

Test cases generated by our approach contain invocation sequences of API functions which we do not usually make because they are obtained by searching reachable states of non-deterministic invocation of those functions. Such test cases allow us to check behavior which is not realized in current applications but will be realized in future ones. This is similar to acceleration testing applied in the field of materials. One can say that we conducted acceleration testing of software in a sense. We think that this testing is important for OSs since they are used in various ways for a long time.

A test case generated can be regarded as an application performed with REL OS. In this sense, we gained evidence to perform a number of applications on the OS. The evidence is important for satisfying safety standards such as IEC61508 [29] and ISO26262 [1].

5.7.4 Creating the Confidence in Correctness

In the verification of seL4 kernel [16, 17], C implementations of the kernel as well as specifications described in Haskell were automatically translated into descriptions of Isabelle/HOL according to pre-defined translation rules. The kernel was verified based not on the implementations themselves but on the translated descriptions. Theorem proving with Isabelle/HOL allows us to create strong confidence in the correctness of proofs done in the verification. On the other hand, a gap between the descriptions of Isabelle/HOL and the implementations still remains. In our approach, the implementation of REL OS was exhaustively executed within specific bounds in testing. We think that executing the implementation itself is very important in order to create the confidence in its correctness. Even though the translated descriptions are verified, it is unimaginable to release the implementations without executing them. Although it is impossible to check all of cases which may happen in the testing, the testing provides evidence that the implementation really works well.

In the existing works on the verification of OSEK OS [19, 21, 23, 24], source codes are only targets to be verified. Since we suspect that the source codes might be incorrect, we need another description which we rely on. In our approach, we rely on the design model which was verified by the model checking. We made much effort to verify the design model so that we could agree that it realized correct behavior of the OS. That is, we created the confidence in the correctness via such design model.

5.8 Conclusion

Applying formal methods to developments of commercial products is often recognized as hard in industry. In fact, we have experience to educate engineers in formal methods [30, 31] and heard such opinions from many of them. In order to persuade them so that formal methods can be practically applied, it is important to show a successful case study of a practical system. In this chapter, we showed a case study that model checking, which is one of formal methods, was applied to an automotive OS. What we should emphasize here is that our target is a commercial product, that is, REL OS. In addition, engineers who develop the product are involved in this project. Showing evidence that model checking has been successfully applied to the commercial product is a primary contribution of this chapter.

We encountered various pragmatical problems in the application of model checking as discussed in the chapter. We solved them by combining engineering techniques such as review and testing. We used informal methods with formal methods for making our approach practical. No bug was found as a result; however, we obtained the confidence that the quality of REL OS is very high. We think that obtaining the confidence is quite different from finding bugs by testing. Clearly, the former is much harder than the latter. The exhaustive techniques that we have adopted allow us to convince that REL OS is correct.

REL OS is going to use in a next series of cars not only for parts which it is currently embedded to but also the other ones. We are convinced that REL OS performs correctly even for the other parts since we have conducted the exhaustive testing which can be regarded as the acceleration testing. The same approach is being applied to the other functions of REL OS. We continue to verify it and extend the approach so that we can acquire more confidence with respect to the quality.

References

1. ISO 26262 Road vehicles—functional safety (2011)
2. Technical Assessment of Toyota Electronic Throttle Control Systems, NHTSA (2011)
3. OSEK/VDX Operating System Specication 2.2.3 (2005)
4. Specification of Operating System 4.0.0, AUTOSAR (2009)

5. L. Sha et al., Priority inheritance protocols: an approach to real-time synchronization. IEEE Trans. Comput. **39**(9), 1175–1185 (1990)

6. G.J. Holzmann, *The Spin Model Checker* (2004)

7. J. Penix et al., Verifying time partitioning in the DEOS scheduling kernel. Formal Methods Syst. Des. **26**(2), 103–135 (2005)

8. M. Dwyer, C. Pasareanu, Filter-based model checking of partial systems, in *Foundations of Software Engineering* (1998), pp. 189–202

9. K. Yatake, T. Aoki, Automatic generation of model checking scripts based on environment modeling, in *International SPIN Workshop on Model Checking of Software* (2010), pp. 58–75

10. Object: Unified Modeling Language: Superstructure, version 2.1.2 (2007)

11. J. Warmer, A. Kleppe, *The Object Constraint Language: Precise Modeling with UML* (Addison-Wesley, Boston, 1999)

12. K. Yatake, T. Aoki, SMT-based enumeration of object graphs from UML class diagrams. ACM SIGSOFT Softw. Eng. Notes **37**(4), 1–8 (2012). International Workshop UML and Formal Methods

13. D. Lee, M. Yannakakis, Principles and methods of testing finite state machines—a survey. Proc. IEEE **84**(8), 1090–1123 (1996)

14. G. Fraser, F. Wotawa, P. Ammann, Testing with model checkers: a survey. J. Softw. Test. Verification Reliab. **19**(3), 215–261 (2009)

15. G. Klein, Operating system verification—an overview. Sādhanā **34**(1), 26–69 (2009)

16. G. Klein et al., seL4: formal verification of an OS kernel, in *ACM Symposium on Operating Systems Principles* (2009), pp. 207–220

17. G. Klein et al., Comprehensive formal verification of an OS microkernel. ACM Trans. Comput. Syst. **32**(1), 1–70 (2014)

18. C. Pasareanu, DEOS kernel: environment modeling using LTL assumptions. NASA ames technical report NASA-ARC-IC-2000-196, NASA Ames Research Center (2000)

19. L. Zhu et al., Formalizing application programming interfaces of the OSEK/VDX operating system specification, in *Theoretical Aspects of Software Engineering* (2011), pp. 27–34

20. E. Cohen et al., VCC: a practical system for verifying concurrent C, in *International Conference on Theorem Proving in Higher Order Logics* (2011), pp. 23–42

21. Y. Huang et al., Modeling and verifying the code-level OSEK/VDX operating system with CSP, in *Theoretical Aspects of Software Engineering* (2011), pp. 142–149

22. PAT, Process Analysis Toolkit 2.9 User Manual. Software Engineering Lab, School of Computing, National University of Singapore (2007)

23. Y. Choi, Safety analysis of trampoline os using model checking: an experience report, in *International Symposium on Software, Reliability Engineering* (2011), pp. 200–209

24. Y. Choi, Model checking trampoline OS: a case study on safety analysis for automotive software. Softw. Test. Verification Reliabil. **24**(1), 38–60 (2014)

25. Trampoline—open source RTOS project, http://trampoline.rtssoftware.org

26. K. Yatake, T. Aoki, Model checking of OSEK/VDX OS design model based on environment modeling, in *International Colloquium on Theoretical Aspect of Computing* (2012), pp. 183–197

27. J. Chen, T. Aoki, Conformance testing for OSEK/VDX operating system using model checking, in *Asia-Pacific Software Engineering Conference* (2011), pp. 274–281

28. J.M. Spivey, The Z notation: a reference manual (1992)

29. IEC 61508: Functional safety of electrical/electronic/programmable electronic safety-related systems (1998)

30. Y. Tahara, N. Yoshioka, K. Taguchi, T. Aoki, S. Honiden, Evolution of a course on model checking for practical applications. ACM SIGCSE Bull. **41**(2), 38–44 (2009)

31. H. Nishihara, K. Shinozaki, K. Hayamizu, T. Aoki, K. Taguchi, F. Kumeno, Model checking education for software engineers in Japan. ACM SIGCSE Bull. **41**(2), 45–50 (2009)

Chapter 6
Formal Methods for Aerospace Systems

Achievements and Challenges

**Marco Bozzano, Harold Bruintjes, Alessandro Cimatti,
Joost-Pieter Katoen, Thomas Noll and Stefano Tonetta**

Abstract The size and complexity of control software in aerospace systems is rapidly increasing, and this development complicates its validation within the context of the overall spacecraft system. Classical validation methods are both labour intensive and error prone as they rely on manual analysis, review and inspection. Thus there is a growing trend to incorporate the use of automated formal methods. This chapter introduces the ESA-funded COMPASS project, which aims at an integrated system-software co-engineering approach focusing on a coherent set of specification and analysis techniques for evaluation of system-level correctness, safety, dependability and performability of on-board computer-based aerospace systems. Its modelling features and supporting toolset provide a unifying framework for system validation, employing state-of-the-art temporal-logic model checking techniques for infinite-state transition systems, both qualitative and probabilistic, with extensions to fault detection, identification and recovery (FDIR) and safety analysis. We provide an overview of the technology and of the results that have been achieved so far, and address several challenges for future developments. Current efforts of the

This work was supported by the European Space Agency through the COMPASS 3 project (ESTEC contract no. 4000115870).

M. Bozzano · A. Cimatti · S. Tonetta
Fondazione Bruno Kessler, Via Sommarive 18, 38123 Povo, Trento, Italy
e-mail: bozzano@fbk.eu

A. Cimatti
e-mail: cimatti@fbk.eu

S. Tonetta
e-mail: tonettas@fbk.eu

H. Bruintjes · J.-P. Katoen · T. Noll (✉)
Software Modeling and Verification Group, RWTH Aachen University, Ahornstraße 55, 52056 Aachen, Germany
e-mail: noll@cs.rwth-aachen.de

J.-P. Katoen
e-mail: katoen@cs.rwth-aachen.de

H. Bruintjes
e-mail: h.bruintjes@cs.rwth-aachen.de

© Springer Nature Singapore Pte Ltd. 2017
S. Nakajima et al. (eds.), *Cyber-Physical System Design from an Architecture Analysis Viewpoint*, DOI 10.1007/978-981-10-4436-6_6

project consortium concentrate on improving and advancing both process as well as technology of the COMPASS approach, with the goal of bringing the methods to higher levels of technology readiness.

Keywords Safety and dependability analysis · Performance analysis · Model checking · AADL modelling language

6.1 Introduction

Verification and validation (V&V) are key processes in the engineering of safety-critical hardware and software systems. Their goal is to check whether the system under construction or its artefacts meet their requirements and its intended functions. The current industry practices for conducting V&V are rather labour intensive [4]. There are severe concerns on scaling these techniques to deal with the ever-growing complexity of systems and in particular of software. The trend is to incorporate the use of formal methods [44, 47, 62]. In particular, *automated verification techniques* are attractive for supporting more rigorous V&V. Formal methods, however, tend to require a high degree of expertise and specialised know-how. These incur substantial investments before their cost and efficiency benefits can be reaped.

To tackle this problem, the European Space Agency (ESA) has initiated an integrated system-software co-engineering approach focusing on a coherent set of specification and analysis techniques for the evaluation of system-level correctness, safety, dependability and performability of on-board computer-based aerospace systems. The work has been and is still being carried out in an ESA-funded project entitled COMPASS, which stands for *COrrectness, Modelling and Performance of AeroSpace Systems* [39].

This chapter gives an overview of the technology and of the results that have been achieved so far, and addresses several challenges for future developments. Current efforts of the project consortium concentrate on improving and advancing *process* as well as *technology* of the COMPASS approach, with the goal of bringing the methods to higher levels of technology readiness.

With regard to technology, several directions to be pursued have been identified, and corresponding methods and implementations are currently under development. Many of them are dealing with failure modelling and analysis. Originally, the COMPASS toolset supports *performability evaluation*: given an AADL model with associated error probabilities, probabilistic model-checking techniques are employed to determine the likelihood of a system failure occurring up to a given deadline. In many cases, however, the probabilities of basic faults are not (exactly) known. It would therefore be worthwhile to consider *parametric* error models, and to automatically compute the maximal tolerable fault probabilities such that the overall model satisfies its performability requirements. A related problem is *model repair*, where one tries to tune the error probabilities of a given model such that a given performability property holds. Moreover, there is increasing demand for verification techniques that are able to cope with several (interdependent) performance measures, such as reliability and energy consumption. In this setting, *multi-objective model checking* is a promising approach.

Another error modelling concept to be investigated further are *Timed Failure Propagation Graphs* (TFPG), which enable a precise description of how and when failures originating in one part of a system affect other parts – a fundamental feature for successfully designing contingency mechanisms. The latter require the careful design and analysis of FDIR strategies, which in turn are based on the automated synthesis of observability requirements to ensure sufficient *diagnosability* of failure situations. Another safety-related concept is *Dynamic Fault Trees*, an expressive extension of standard Fault Trees that additionally cater for common dependability patterns. In COMPASS, their analysis relies on the extraction of an underlying stochastic model, which is a time-consuming process. This can be improved by reducing the size of this model prior to analysis using graph transformation techniques, and by accelerating the state space generation by leveraging reduction techniques from model checking.

Another direction which is currently under development, is the enhancement of the tool support to *formalise* the requirements into formal properties and to *validate* with formal techniques that the specification is correct and complete. Related to this, the specification of formal properties in terms of component assumptions and guarantees enables *contract-based design*, including the verification of contract-based refinement and contract-based compositional verification of the system behaviour.

The following section sketches a systematic space systems engineering approach as advocated by ESA and related institutions. In the subsequent two sections, we address both the results that have been achieved in the COMPASS project and some of the remaining challenges, and then conclude with a brief summary.

6.2 Space Systems Engineering

ESA and related institutions develop and maintain a series of standards for the management, engineering and product assurance in space projects and applications, known as European Cooperation for Space Standardization (ECSS). Among others, Standard ECSS-E-ST-10C [48] specifies the system engineering implementation requirements for space systems and space products development. More concretely, it states that

Systems engineering is defined as an interdisciplinary approach governing the total technical effort to transform a requirement into a system solution. A system is defined as an integrated set of elements to accomplish a defined objective. These elements include hardware, software, firmware, human resources, information, techniques, facilities services, and other support elements.

Moreover, [48] partitions system engineering into the following activities:

requirements engineering   which consists of requirement analysis and validation, requirement allocation, and requirement maintenance;

analysis   which is performed for the purpose of resolving requirements conflicts, decomposing and allocating requirements during functional analysis, assessing system effectiveness (including analysing risk factors); and complementing testing evaluation and providing trade studies for assessing effectiveness, risk, cost and planning;

design and configuration   which results in a physical architecture, and its complete system functional, physical and software characteristics;

verification   whose objective is to demonstrate that the deliverables conform to the specified requirements, including qualification and acceptance;

system engineering integration and control   which ensures the integration of the various engineering disciplines and participants throughout all the project phases.

The following section describes to what extent these activities are supported in our COMPASS approach.

6.3   Achievements

The COMPASS project funded by the European Space Agency (ESA) aims at an integrated system-software co-engineering approach focusing on a coherent set of specification and analysis techniques for evaluation of system-level correctness, safety, dependability and performability of on-board computer-based aerospace systems. Its main contributions are a tailored modelling language and a toolset for supporting (semi-)automated validation activities. The modelling language is a dialect of the Architecture Analysis and Design Language (AADL) and enables engineers to specify the system, the software, and their reliability aspects. The COMPASS toolset provides a unifying framework for validation, employing state-of-the-art temporal-logic model checking techniques for infinite-state transition systems, both qualitative and probabilistic, with extensions to fault detection, identification and recovery (FDIR) and safety analysis. Its applicability has been demonstrated in several case studies in the space domain, ranging from thermal regulation and mode management in satellites with associated FDIR strategies to an industrial-size satellite platform. Here we provide a brief overview of our framework. A more comprehensive description is given in [24, 65, 72].

6.3.1   The COMPASS Approach

The COMPASS toolset addresses, in a coherent manner, different aspects that are relevant to the engineering of complex systems, such as co-engineering of hardware and software, performability and dependability, reliability, availability, maintainability and safety engineering (RAMS). COMPASS offers a multi-disciplinary approach that supports the early design phases by considering systems at the architecture level. Thus it mainly targets the "requirements engineering" and "analysis" functions of system engineering, but also tackles the "design and configuration" and "verification" phases.

More concretely, COMPASS provides a specification language that offers convenient ways to describe nominal hardware and software operation, hybridity, (probabilistic) faults and their propagation, error recovery, and degraded modes of operation. This language is discussed in Sect. 6.3.2 in greater detail. It is equipped with a formal semantics that opens up the possibility to apply a wealth of formal methods for various kinds of verification and validation activities. Most of these are based on formal requirements as introduced in Sect. 6.3.3. V&V is supported by an integrated toolset, described in Sect. 6.3.4, that covers the following functionalities according to the ECSS standards.

Requirements Validation [48] In order to ensure the quality of requirements, they can be validated independently of the system. This includes both property consistency (i.e., checking that requirements do not exclude each other), property assertion (i.e., checking whether an assertion is a logical consequence of the requirements), and property possibility (i.e., checking whether a possibility is logically compatible with the requirements). Altogether these features allow the designer to explore the strictness and adequacy of the requirements. Expected benefits of this approach include traceability of the requirements and easier sharing between different actors involved in system design and safety assessment. Furthermore, high-quality requirements facilitate incremental system development and assessment, reuse and design change, and they can be useful for product certification.

Functional Verification [48] Analysing operational correctness is the first step to be performed during the system development lifecycle. It consists in verifying that the system will operate correctly with respect to a set of functional requirements, under the hypothesis of nominal conditions, that is, when software and hardware components are assumed to be fault-free. This can be accomplished by both simulation and exhaustive model checking techniques.

Safety and Dependability Analysis [49, 51–53] Analysing system safety and dependability is a fundamental step that is performed in parallel with system design and verification of functional correctness. The goal is to investigate the behaviour of a system in degraded conditions (that is, when some parts of the system are not working properly, due to malfunctions) and to ensure that the system meets the safety requirements that are required for its deployment and use.

Performability Analysis [50] To guarantee the required system performance in the presence of faults, integrated hardware and software models can be evaluated with respect to their performance behaviour in degraded modes of operation. In line with the approach for the functional correctness, again model checking techniques are employed for assessing this type of requirements.

Fault Detection, Identification and Recovery (FDIR) Analysis [49] System models can include a formal description of both the fault detection and identification sub-systems, and the recovery actions to be taken. Based on these models, tool facilities are provided to analyse the operational effectiveness of FDIR measures, and to investigate the observability requirements that make the system diagnosable.

In summary, the overall process of analysing system specifications in the COMPASS framework involves the following steps:

1. System specifications (describing the nominal and, if applicable, the error behaviour) are entered using a text editor, and loaded into the toolset (cf. Sect. 6.3.2).
2. Some of the subsequent analyses require writing properties. COMPASS offers several ways to specify such properties (cf. Sect. 6.3.3).
3. To interactively explore the dynamic behaviour of the system, the model simulation feature of the toolset can be employed (cf. Sect. 6.3.4).
4. Finally, depending on the type of the system, a plethora of analyses can be applied (cf. Sect. 6.3.4).

6.3.2 System Modelling

AADL [84] is an industry standard for modelling safety-critical system architectures, which is developed and governed by SAE. This language provides a cohesive and uniform approach for modelling heterogeneous systems, consisting of software (e.g., processes and threads) and hardware (e.g., processors and buses) components, and their interactions. It enables analysis of system designs prior to implementation and supports a model-based and model-driven development approach throughout the system life cycle.

Our dialect of AADL was designed to meet the needs of the European space industry. The original language is mainly focused on the architectural organisation of a system under nominal and degraded modes of operation. The nominal modes indicate that the system is operating normally, whereas degraded modes typically signify that the system's functions are (partially) impaired due to some anomaly. Our goal was to extend AADL's scope on defining the architecture of a system by also allowing to analyse its dynamic behaviour, namely both its nominal and degraded modes of operation and their interweaving.

In particular, quantitative aspects such as the timing of operations and the likelihood of faults should be covered. To this end, we built on a core fragment of AADL Version 1 [83] and extended it, essentially by supporting the following features:

- Modelling both the system's *nominal* and *faulty* behaviour. To this aim, primitives are provided to describe software and hardware faults, error propagation (i.e., turning fault occurrences into failure events), sporadic (transient) and permanent faults, and degraded modes of operation (by mapping failures from architectural to service level).

- Modelling (*partial*) *observability* and the associated *observability requirements*. These notions are essential to analyse the effectiveness of fault management systems. These subsystems, being part of the overall system, monitor it, identifying when a fault has occurred, pinpointing the type of fault and its location, and finally recovering from it by, for example, switching to a backup system configuration.
- Specifying *timed* and *hybrid* behaviour. In particular, to analyse physical systems with non-discrete behaviour, such as mechanics and hydraulics, the modelling language supports continuous real-valued variables with (linear) time-dependent dynamics.
- Modelling *probabilistic* aspects, such as random faults and repairs, that are subject to stochastic timing.

A complete system specification consists of three parts, namely a description of the nominal behaviour, a description of the error behaviour, and a fault injection specification that describes how the error behaviour influences the nominal behaviour. This separation approach is different from the one taken in AADL and its Error Model Annex [81], which interacts through an explicit specification of mangling error and nominal events. In contrast, our dialect provides an automated mechanism (called *model extension*) that enables engineers to keep the nominal model completely separate from the error model. A comprehensive presentation of our specification language and its formal semantics is given in [25].

6.3.3 Requirements Specification

An important aspect of V&V of requirements is the consistent and complete specification of formal properties associated with the requirements. As the (correct) specification of such properties requires a significant amount of technical expertise, the COMPASS toolset has striven to alleviate this burden from its users as much as possible.

The approach initially taken by COMPASS was to allow the user to specify properties by means of *patterns* [5, 46]. These provide a structured way of generating a formal property given a template with placeholders, where the user provides basic propositions (statements about the current state of the system) for each of these placeholders. The use of these patterns relies on the fact that the requirements themselves often use recurring shapes. For example, one pattern describes the *absence* of particular behaviour in the system, e.g., reaching a state of critical failure.

Various logics are supported by this approach, in particular qualitative, timed and probabilistic logics, such as LTL/CTL, MITL and CSL respectively. A formal property expressed by a pattern is converted into the appropriate logic depending on the pattern used, and given to a model checker for analysis. An appropriate model checker in the toolset will be provided the input model and one or more properties, and checks whether the property holds, thus providing the formal verification of the requirement associated with it.

Recently, support has been added to COMPASS for the *Catalogue of System and Software Properties* (CSSP) in the CATSY project [18]. The CSSP defines a set of design attributes which are used to automatically derive formal properties. A requirements taxonomy has been set up, with a focus on the space engineering domain, to allow various project requirements to be classified (e.g., requirements related to monitoring, protocols or availability). For each of these classes, associated design attributes have been identified. Such design attributes may refer to the presence of elements in the model (e.g., redundant components or mode transitions) or to properties associated with such elements (e.g., timing of events, or reactions to events). The latter are collected in the CSSP. This way, for appropriate requirement classes, a user can simply specify the value of the associated properties. A formal property is generated automatically for such model properties, allowing for the verification of the corresponding requirement.

Moreover, still within the CATSY project, the specification language has been enriched with the possibility to specify properties on components directly in temporal logics, thus without the help and limitation of the pattern-based approach. Finally, the properties attached to components (either specified directly in temporal logic or by means of patterns or the CSSP) can be structured into *contracts* [38], i.e., pairs of assumptions and guarantees, to enable the verification of contract-based refinement and contract-based compositional verification and safety analysis.

6.3.4 COMPASS Toolset

The COMPASS toolset is the result of a significant implementation effort carried out by the COMPASS Consortium. The GUI and most subcomponents are implemented in Python, using the PyGTK library. Pre-existing components, such as the NuSMV and MRMC model checkers, are mostly written in C. Overall, the core of the toolset consists of about 100,000 lines of Python code. Figure 6.1 shows the functionality of the toolset. Its main features are introduced in the COMPASS Tutorial [41]. It is complemented by the COMPASS User Manual [40], which can be consulted for a more systematic reference.

COMPASS takes as input one or more AADL models, and a set of properties. As pointed out before, the latter are provided in the form of generic properties or instantiated property patterns, which are templates containing placeholders that have to be filled in by the user. The COMPASS toolset provides templates for the most frequently used patterns, that ease property specifications by non-experts through hiding the details of the underlying temporal logic. The tool generates several outputs, such as traces, Fault Trees and FMEA tables, diagnosability and performability measures.

The toolset builds upon the following main components. NuSMV (New Symbolic Model Verifier, [34, 74]) is a symbolic model checker that supports state-of-the-art verification techniques such as BDD-based and SAT-based verification for CTL and LTL [8]. nuXmv [32] is an extension of NuSMV for the SMT-based verification of

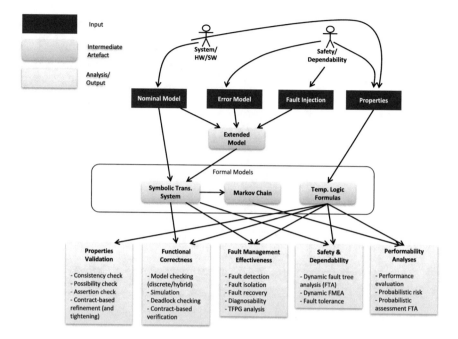

Fig. 6.1 Functional view of the COMPASS platform

infinite-state systems. MRMC (Markov Reward Model Checker, [67, 71]) is a probabilistic model checker that enables the analysis of discrete-time and continuous-
time Markov reward models. Specifications are written in PCTL (Probabilistic
Computation Tree Logic) and CSL (Continuous Stochastic Logic [6], a probabilistic real-time version of CTL). SigRef [88] is used to minimise, amongst others,
Interactive Markov Chains (IMC; [61]) based on various notions of bisimulation.
It is a symbolic tool using multi-terminal BDD representations of IMCs and applies
signature-based minimisation algorithms. xSAP [13] is a tool that supports model-
based safety analysis including Fault-Tree Analysis, FMEA, and diagnosability.
OCRA [36] takes in input a system architecture specification and allows to perform
contract-based validation and verification.

The tool also supports a graphical notation of our AADL dialect, which is derived
from the graphical notation of AADL [82]. We developed a graphical drawing editor
enabling engineers to construct models visually using the adopted graphical notation.
This editor is called the *COMPASS Graphical Modeller* and is part of the COMPASS
toolset. Figure 6.2 shows the main window of the COMPASS toolset after loading a
system model and performing a fault injection.

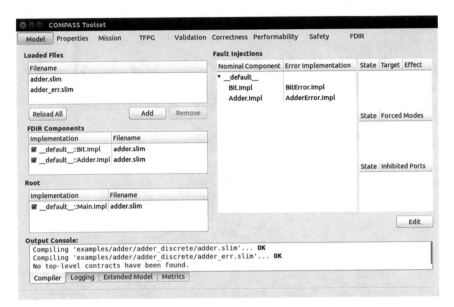

Fig. 6.2 Main window of the COMPASS toolset

6.3.4.1   Properties Validation

Figure 6.3 gives an example of a property specification by means of patterns. Before analysing the correctness of the behavioural model against the specified properties, the properties themselves can be validated to search for errors in the requirements or in their formalisation. COMPASS supports this activity of properties assurance [79] by allowing the user to specify and check *property validation problems*. These can be a simple check of consistency (i.e., logical satisfiability) of a set of properties or can consist of specifying a new property to be consistent with or entailed by a set of existing properties. In case of inconsistency or failed entailment, an execution trace is generated as a witness of the result. In case of proved inconsistency or entailment, a minimal subset of properties that are sufficient for the proof can be extracted.

When properties are structured into contracts, COMPASS supports the verification of their refinement as described in [38]. In *contract-based design*, the assumptions of a component are properties to be satisfied by the component environment, while the guarantees are properties to be satisfied by the implementation when the assumptions hold. A correct *contract refinement* ensures that any correct implementation of the subcomponents form a correct implementation of the composite component, and, together with an environment satisfying the assumptions, form a correct environment for each subcomponent. This is verified by generating and proving a set of proof obligations, which are validity problems for the underlying temporal logic. For every refined contract, there is a proof obligation to ensure that the guarantee of the composite component is entailed by the conjunction of the assumption of the

Fig. 6.3 The property editor

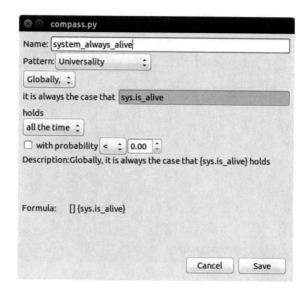

composite component and the contracts of the subcomponents, and similarly for each assumption of each subcomponent.

If the contract refinement is not correct, the tool provides an execution trace for every invalid proof obligation. In case the refinement is correct, the tool can provide some feedback by presenting viable *tightenings* of the contract refinement, i.e. stronger/weaker versions of the assumptions/guarantees that still yield a correct contract refinement, as described in [35].

6.3.4.2 Functional Correctness

COMPASS supports random and guided *model-based simulation* of AADL models. Guided simulation can be performed by choosing either the next transition to be taken, or a target value for one or more variables. The generated traces can be inspected using a trace manager that displays the values of the model variables of interest (filtering is possible) for each step.

Property verification is based on model checking [8], an automated technique that verifies whether a property expressed in temporal logic holds for a given model. Symbolic techniques [10, 11, 60] are used to tackle the problem of state space explosion. COMPASS relies on the NuSMV [34, 74] and nuXmv [32, 75] model checkers, which support both BDD-based and SAT-based verification for finite-state systems, and SMT-based verification techniques for timed and hybrid systems, based on the MathSAT solver [20, 69]. On refutation of a property, a counterexample is generated, showing an execution trace of the model violating the property. An example of this is shown in Fig. 6.4. It is also possible to run *deadlock checking*, in order to pinpoint deadlocks (i.e., states with no outgoing transitions) in the model.

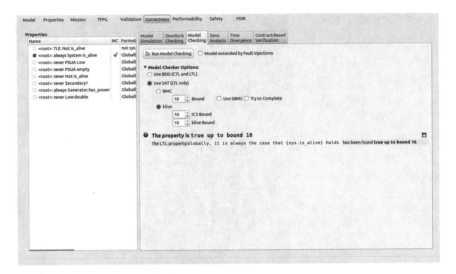

Fig. 6.4 A model-checking counterexample

The verification of properties can be enhanced by contract-based specification to perform it in a compositional way. In this case, COMPASS interacts with OCRA [36], first, to check that the contract refinement is correct and, second, to individually verify each atomic component with respect to its local contract. These checks ensure the correctness of the overall system by compositional reasoning.

6.3.4.3 Safety and Dependability Assessment

COMPASS implements model-based safety assessment techniques, based on symbolic model checking [26, 28], and supports traditional techniques such as *Failure Mode and Effects Analysis* (FMEA; [49]) and *Fault Tree Analysis* (FTA; [52]). FMEA is an inductive technique that starts by identifying a set of (combinations of) failure modes and, using forward reasoning, assesses their impact on a set of system properties. The results are summarised in an *FMEA table*. It is also possible to generate *dynamic* FMEA tables, i.e., to enforce an order of occurrence between failure modes. FTA is a deductive technique, which, given a *top-level event* (TLE), i.e., the specification of an undesired condition, constructs all possible chains of basic faults that contribute to its occurrence. Pictorially, these chains are organised in a *Fault Tree* with a two-layer logical structure, corresponding to the disjunction of its minimal cut sets (MCSs; [28]), where each MCS is a conjunction of basic faults. COMPASS also supports the generation of (a subset of) *Dynamic Fault Trees* [45], where ordering constraints between basic faults are represented using priority AND (PAND) gates. Figure 6.5 depicts a Fault Tree.

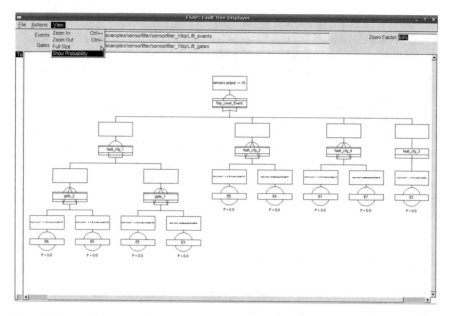

Fig. 6.5 A generated fault tree

It is also possible to exploit a contract-based specification to obtain a fault tree that follows the hierarchical decomposition of the system [27]. In this case, COMPASS interacts with OCRA to compute a fault tree where each intermediate event represents the failure to satisfy either the guarantee or the assumption of a contract. The fault tree represents the dependency between such failure and the failures of other components. For example, the failure of a composite component can be caused by the combined failure of two of its subcomponents or by the failure of its assumption, which in turn can be caused by the failure of other components. Compared to the "monolithic" FTA described above, the contract-based FTA is more pessimistic with regard to the identification of possible system failures because it follows the conservative approximation given by the contract-based refinement. The resulting fault tree is however often more intuitive because it uses intermediate events corresponding to the components in the system architecture.

6.3.4.4 Diagnosability and FDIR Analysis

The COMPASS toolset supports diagnosability and FDIR (Fault Detection, Isolation and Recovery) effectiveness analysis. These analyses work under the hypothesis of *partial observability*. Variables and ports in our AADL dialect can be declared to be observable.

Diagnosability analysis investigates the possibility for an ideal diagnosis system to infer accurate and sufficient run-time information on the behaviour of the observed system. The COMPASS toolset follows the approach described in [37], where the violation of a diagnosability condition is reduced to the search of *critical pairs* in the so-called *twin plant* model, i.e., pairs of execution traces that are observationally equivalent but hide conditions that should be distinguished. Figure 6.6 shows such a pair of traces.

FDIR effectiveness analysis refers to a set of analyses carried out on an existing fault management subsystem. *Fault detection* is concerned with detecting whether a given system is malfunctioning, namely searching for observable signals such that every occurrence of the fault will eventually trigger them. *Fault isolation analysis* aims at identifying the specific cause of malfunctioning. It generates a Fault Tree that contains the minimal explanations that are compatible with the observable being triggered. Finally, *fault recovery analysis* is used to check whether a user-specified recoverability property holds.

6.3.4.5 Timed Failure Propagation Graphs

COMPASS supports *Timed Failure Propagation Graphs* (TFPGs; [1, 70, 76]) as a means to model and analyse how failures originating in one part of a system affect other parts. Traditionally, TFPGs can be used for both diagnosis and prognosis. TFPGs describe the occurrence of failures and the temporal interrelationships between failures and their direct and indirect effects. They constitute a very rich formalism that can express Boolean combinations of basic failures, intermediate consequences, and transitions across them, labelled with propagation times and possibly dependent on the system's operational modes. TFPGs are increasingly used for

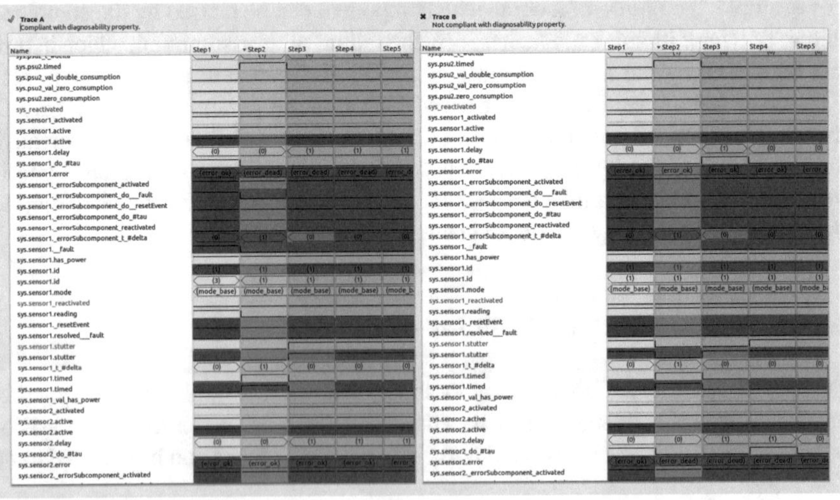

Fig. 6.6 Diagnosability counterexample

the design of autonomous systems, in particular for the design of FDIR procedures. Compared to other techniques such as FTA and FMEA, TFPGs have substantial advantages. They present a more comprehensive and integrated picture than Fault Trees, as they focus on propagation paths in response to individual feared events. Moreover, in comparison to FMEA tables they provide additional and more precise information, such as timing information and AND/OR correlations between propagation causes and effects.

As shown in Fig. 6.7, COMPASS enables the modelling and analysis of TFPGs. The available analyses include *behavioural validation*, that is, verification that a TFPG is a complete representation of failure propagation with respect to a given system model, and *effectiveness validation*, that is, verification that a TFPG is a suitable model for diagnosis, i.e., contains sufficient information to carry out diagnosis, discriminating between different possible causes. Finally, COMPASS supports the automatic synthesis of a TFPG, given a set of nodes and a system model.

6.3.4.6   Performability Analysis

We use *probabilistic model checking* techniques [7, 8] for analysing a model with respect to its performance. The COMPASS toolset in particular supports performance properties expressed by the probabilistic pattern system presented in [5]. It allows for the formal specification of steady-state, transient probabilities, timed reachability probabilities and more intricate performance measures such as combinations thereof. An example of a typical performance parameter is "the probability that the first battery dies within 100 h" or "the probability that both batteries die within the mission

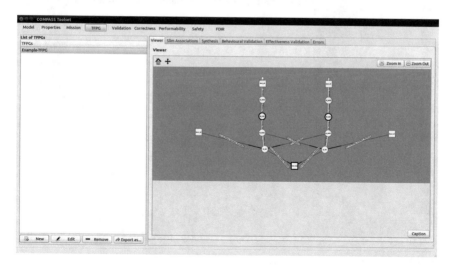

Fig. 6.7 Example of a TFPG

duration". These properties have a direct mapping to Continuous Stochastic Logic (CSL; [6]) and are input to the underlying probabilistic model checker.

The probabilistic model checker furthermore requires a Markov model as input. This is obtained from the integrated nominal and error model through several steps. First, the extended model's reachable state space is generated through an exhaustive symbolic exploration. Second, the probabilistic rates as specified in the error models are interwoven through the state space by replacing the transition label with the associated probabilistic rate. The resulting state space is a symbolic representation of an Interactive Markov Chain (IMC), i.e., a Continuous-Time Markov Chain (CTMC) that may exhibit non-determinism [61]. This IMC is passed through the third phase, in which its size is reduced using weak bisimulation minimisation [43, 86]. In the final phase, CSL formulae are extracted from the performance requirements, and are fed together with the reduced IMC to a probabilistic model checker, to compute the desired probabilities. If the reduced IMC is a proper CTMC, the MRMC model checker [67] is used for this purpose, otherwise the IMCA model checker [58] is employed. As can be seen in Fig. 6.8, the result is a graph showing the cumulative distribution function over the time horizon specified in the performance requirement. Similar techniques are also used for *Fault Tree evaluation*, i.e., for computing the probability of the top-level event in Dynamic Fault Trees [19].

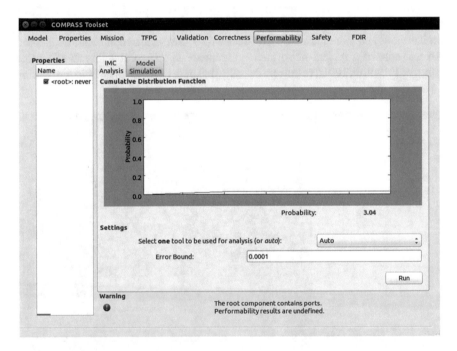

Fig. 6.8 Performing performability analysis

6.3.5 Case Studies

The COMPASS methodology has been progressively assessed by several industrial case studies, steadily increasing in size and scope. Table 6.1 summarises the results, respectively giving appropriate references, the number of components of the system model ("#C"), the main aspect to be investigated, and the major technological challenges that have been solved and those that were left open (and will be addressed in Sect. 6.4).

Thales Alenia Space conducted the first evaluation. They developed two case studies of their satellite subsystems, respectively dealing with mode management and thermal regulation, and analysed them using the COMPASS toolset [65]. These subsystem case studies demonstrated the potential of understanding the subtle interactions between the system, software, and the fault management system. They furthermore raised follow-up questions: how would models with a greater level of detail be handled? In which phases of the systems engineering life cycle is the COMPASS methodology particularly suitable?

To address these issues, ESA conducted a laboratory project to model a full satellite platform using the COMPASS methodology. This was performed at phase B of the space systems engineering life cycle, the preliminary system design [24, 54]. A subsequent laboratory project was initiated afterwards to model a full satellite platform at phase C, the detailed system design. Here, special focus was put on diagnosability analysis, which in the phase B pilot was deemed intractable. This analysis increasingly gains importance in the engineering life-cycle as fault management designs become more involved to meet mission demands. Our experiences indicate a clear need for enhanced diagnosability analysis algorithms that also account for delayed diagnostic means. The outcomes of this study are discussed in detail in [24].

Moreover, [31] presents a case study performed together with Airbus Defense and Space, which was carried out to demonstrate the applicability of stochastic model checking (Monte Carlo methods) to analyse timed reachability properties of a simplified launcher system. The evaluation revealed the need to support more expressive kinds of stochastic logics.

Recently, a case study on the application of TFPGs for the on-going Solar Orbiter project [85] was performed internally at ESA. It shows the feasibility of using TFPG-

Table 6.1 Overview of case studies

Case study		#C	Aspect	Solved	Open
Mode mgmt.	[65]	3	Fault mgmt.	Scalab. of analysis	—
Thermal reg.	[65]	12	Hybrid behav.	Zeno paths	—
Platform B	[24, 54]	86	FDIR	Fairness	—
Platform C	[24]	246	Diagnosab.	Effic. diagnosab. anal.	Delayed diagnosis
Launcher	[31]	37	Prob. reachab.	Effic. performab. anal.	Expressivity of logics
Solar Orbiter	[12]	15–39	Failure propag.	TFPG analysis	Fault recov. synthesis

based analyses to study time-critical failure propagation at a unit to subsystem level. When using focused modelling, also the analysis of timed failure propagation in detailed physical models is feasible. In general, TFPGs have been found to be a promising technology to formally integrate various key aspects of FDIR design, including discrete failure propagations across all levels of a system, time bounds on the delays, mode constraints, and monitors. This is very important, because informal analysis usually done with FMEA makes it difficult to demonstrate completeness and timing properties of the proposed design. Indeed, by application of TFPGs in the case study one case of a propagation link missing in the FMEA tables has been identified.

6.4 Challenges

With regard to technological challenges, several directions to be pursued have been identified, and corresponding methods and implementations are currently under development.

6.4.1 Formal Validation of Probabilistic Properties

As explained in Sect. 6.3.3, system requirements are formalised by temporal logics. The toolset described in Sect. 6.3.4 supports the validation of such properties by means of satisfiability/validity checking, i.e., the problem of deciding whether a given formula is satisfied by at least one or, dually, by every system model. This is currently supported for qualitative logics like LTL and CTL, which only allow to describe the order of system actions but cannot express quantitative properties such as timing or probabilities. These logics possess the so-called finite model property, and the complexity of checking satisfiability has been explored for various fragments.

However, the satisfiability problem for probabilistic versions of CTL such as PCTL and CSL, which are employed to express quantitative properties of system models, is almost unexplored [33]. These logics are quite popular in the field of probabilistic verification as their model-checking problem is known to be decidable. When it comes to satisfiability, however, the analysis turns out to be a much more difficult endeavour. In fact, this is a long-standing open problem for PCTL. Results so far are restricted to logical fragments such as qualitative PCTL [29], or are obtained by considering variations of the satisfiability problem. One of the most recent results is given in [33], where the satisfiability problem for a bounded fragment of PCTL is shown to be decidable.

Therefore, there is strong demand for identifying richer fragments of PCTL and for studying their satisfiability, complexity, and finite/rational model properties. A promising direction is to start from the characterisation of safety and liveness fragments of this logic [66].

6.4.2 Contract-Based Fault Injection

As discussed in Sect. 6.3.4, Fault Tree Analysis can be performed either by means of a more traditional model-based approach, which computes the minimal cut set for a top-level event, or by means of contract-based safety analysis, which produces a hierarchical Fault Tree that follows the specified contract refinement (and thus the architectural decomposition). The two analyses are currently disconnected: while the model-based safety analysis exploits the error model specification to automatically inject faulty behaviours into the nominal model, the contract-based approach identifies a failure by the fact that the component implementation violates a guarantee or that the component environment does not satisfy an assumption. Thus, in the contract-based approach, in case of failure, any behaviour is possible. This may result in Fault Trees that describe combinations of failures which can never occur in the real system. An interesting research direction is to find an effective way to inject the faults in the contract specification in order to have degraded assumptions and guarantees in case of failures.

6.4.3 FDIR Design and Diagnosability

The area of diagnosability and fault detection, identification [and recovery] (FDI[R]) design is particularly challenging. Recent work [22, 23] has addressed the extension of diagnosis and FDI to incorporate the notion of *delay*, and to address cases where diagnosability cannot always be guaranteed for all system executions.

The diagnosis delay characterises situations where diagnosis requires a time delay in order to be carried out. The notion of *alarm condition* formalises the relation between the condition to be diagnosed (e.g., presence or absence of a fault, isolation between different faults) and the raising of an alarm by the diagnoser; an alarm condition may specify diagnosis with an exact delay (after exactly a given time), a bounded delay (within a given time) and finite delay (eventually). Moreover, the notion of *trace diagnosability* formalises cases where diagnosability cannot be guaranteed globally, but only locally, on a subset of traces, and the notion of *maximality* formalises the capability of a diagnoser to raise the alarm as soon as possible and as long as possible. Finally, FDI effectiveness properties state the correctness and completeness of an FDI design with respect to the FDI requirements. In [23], all of these properties of an FDI design can be specified using a general framework and language based on temporal epistemic logic. Verification can be performed using an epistemic model checker. Alternatively, the diagnosability check can be reduced to standard temporal logic model checking based on the twin-plant approach described in Sect. 6.3.4. The original version of the latter was introduced in [37] and is being employed in the current COMPASS toolset, while [22, 57] shows how to extend it to deal with epistemic logic. In addition, an algorithm for the automatic synthesis of correct-by-construction FDI components is available [23].

Another interesting research area concerns the analysis of observability requirements for diagnosis, and the synthesis of a set of observables that are sufficient to
ensure diagnosability [16]. It is possible to rank configurations of observables based
on cost, minimality, and diagnosability delay, thus helping designers in finding the
most appropriate configuration.

The automatic *synthesis* of FDIR components has been considered in two projects
related to COMPASS, namely AUTOGEF [2] and FAME [15], also funded by ESA.
The problem is cast in the frame of discrete event systems and finite delay diagnosis,
and is tackled by synthesizing the fault detection and fault recovery components
separately, with the idea that fault recovery implements a plan (recovery strategy) to
respond to the alarms generated by the fault detection component. FAME addresses
the problem of FDIR synthesis for continuous-time systems, where the diagnoser
communicates with the plant by sampling the values of the sensors at periodic time
intervals. Another outcome of the FAME project is the definition of a general process
for FDIR design that spans the different phases of system development (definition of
mission and FDIR requirements, safety assessment, FDIR design and verification).
This process aims at enabling a consistent and timely FDIR conception, development,
verification and validation, overcoming several shortcomings of existing practices.
FDIR synthesis, along with other functionality described in this section, is under
consideration for inclusion in the COMPASS toolset.

6.4.4 Timed Failure Propagation Graphs

Recent work has focused on techniques for validating TFPG models. In particular,
[21] studies several validation problems using advanced techniques based on satisfiability modulo theory, namely possibility and necessity, refinement and diagnosability.
Moreover, [17] addresses both the completeness of a TFPG with respect to a system model, and the problem of *tightness* of TFPG edges, that is, the possibility that
certain parameters, specifically time bounds, of the TFPG can be reduced without
breaking its completeness. Finally, the problem of automatic synthesis of a TFPG is
thoroughly investigated in [14]. Automatic tightening of the TFPG nodes, coupled
with the synthesis of the graph, may be used to automatically produce a complete
and tight TFPG from a system model, given the definition of the TFPG nodes.

6.4.5 Parametric Error Models

Originally, the COMPASS toolset supports performability evaluation (cf. Sect. 6.3.4):
given an AADL model with associated error probabilities, the likelihood of a system
failure occurring up to a given deadline is determined. The underlying technique
is probabilistic model-checking. In many cases, however, the probabilities of basic
faults are not known, or at best can be estimated by lower and upper bounds. It

would therefore be worthwhile to consider parametric error models, in which the probabilities of faults are (partially) left open.

Parameter synthesis focuses on automatically computing the maximal tolerable parameter values such that the resulting model satisfies its performability requirements. Although this problem is inherently harder than (probabilistic) model checking, first results indicate that for a limited number of parameters, solutions are feasible and scalable [42, 63]. They would allow to derive quality requirements for electronic [80] or mechanical parts [73] or software components [3] of a system to be developed.

A related problem is *model repair*, where one tries to tune the error probabilities of a given model such that the resulting model satisfies a given performability requirement. Current approaches only consider changes of the transition probabilities, whereas modifications of the underlying topological structure are not considered. Different methods exist, such as global repair [9] and the more recent technique to perform local repair operations in an iterative fashion [77].

6.4.6   Dynamic Fault Trees

Fault Tree Analysis (cf. Sect. 6.3.4) is a widespread industry standard for assessing system reliability [52]. Standard (static) Fault Trees model the failure behaviour of a system dependent on its component failures. To overcome their limited expressive power, *Dynamic Fault Trees* (DFT) have been introduced to model advanced dependability patterns, such as spare management, functional dependencies, and sequencing [45]. Currently, in addition to static Fault Trees the COMPASS toolset only supports sequencing, by representing ordering constraints between basic faults using priority AND (PAND) gates. However, there is strong demand for improving safety assessment by supporting more expressive constructs in DFTs. They often lead to fault models that are more succinct, and thus better comprehensible.

DFT analysis relies on the extraction of an underlying stochastic model, such as a Bayesian Network, a Continuous-Time Markov Chain, a Stochastic Petri Net, or an Interactive Markov Chain. This is a time-consuming process, in particular for more expressive dependency patterns, raising the need for approaches to make it simpler and cheaper (in terms of computational resources). A key technique is the *reduction of the state space* of DFTs prior to (and during) their analysis. Here, one technique is to consider DFTs as (typed) directed graphs and to manipulate them by graph transformation, a powerful technique to rewrite graphs via pattern matching. In [64], a catalogue of 28 (templates of) rules is presented that convert a given DFT into a smaller, equivalent one having the same system reliability and availability characteristics. Experiments with 170 DFTs, originating from standard examples from the literature as well as industrial case studies from aerospace and railway engineering, showed encouraging results. The rewriting approach enabled us to cope with 49 DFTs that could not be handled before. But also for static Fault Trees the processing pays off, rendering analysis much faster and more memory efficient, up to two orders of magnitude.

More state-space reductions can be obtained by tailoring two successful techniques from the field of model checking, namely, symmetry reduction and partial-order reduction [87]. In the DFT setting, this amounts to the detection of isomorphic sub-DFTs, of stochastic independencies, and of sub-DFTs that become obsolete after the occurrence of some faults. In addition, certain failure orderings arising from superfluous non-determinism can be ignored in the analysis. All this comes at no run-time penalty: as the results in [64, 87] indicate, structural transformations of DFTs operate very fast, and the stochastic model generation is significantly accelerated due to the reduction. This opens the possibility of supporting more expressive types of Fault Trees, and considering techniques on how to analyse them efficiently in the COMPASS toolset.

In addition, one can exploit that some stochastic analyses such as assessing the reliability of a system—how likely is it operational up to a certain point in time?—are *compositional*: the measure can directly be computed from its sub-DFTs' measures. This means that the analysis can be carried out in a modular way by considering only a part of the state space in each step, and by re-computing measures incrementally after local changes in DFTs.

Last but not least, the *parameter synthesis* techniques as sketched in Sect. 6.4.5 can also be applied to DFTs. Classical analyses require all component failure rates to be known, which often does not hold in practice. Thus, a relevant problem is to synthesise the allowed component failure rates ensuring, e.g., a given mean minimal time between failures. This is clearly an instance of the parameter synthesis problem as described earlier.

6.4.7 Multi-objective Verification

Besides the correctness of their functional behaviour, systems are required to exhibit adequate *performance* characteristics. The latter can be measured by, e.g., its average and peak energy consumption, construction costs, and its availability and reliability. These measures are often contradictory: while using more power for data transmission typically increases the reliability level of communication, it also entails a higher energy consumption. But also less obvious mutual dependencies can emerge: optimising a system for (long-run) availability might reduce the (short-term) reliability.

In order to systematically investigate such effects, *multi-objective model checking* can be employed. This is a fully automatic technique by which, based on a model of the system under consideration and some measures-of-interest, a so-called Pareto curve is deduced [56]. The latter gives an (often graphical) representation of the optimal strategy for resolving non-deterministic choices in the system with respect to a given weighting of these measures. Currently, only Markov Decision Processes can be handled by this technique [55, 68]. While these support non-deterministic choices (to be optimised) and discrete probabilities, they lack continuously distributed random delays, which are typically used to describe, e.g., mechanical wear or other sources of failures.

Markov Automata [59] constitute a highly expressive formalism which extends Markov Decision Processes by such random delays. They are known to provide a suitable model for, e.g., Dynamic Fault Trees (cf. Sect. 6.4.6) or to define formal semantics of Stochastic Petri Nets. Previous work is only able to cope with optimising single measures on Markov Automata [59]. Our aim is therefore to extend the techniques that have been developed for Markov Decision Processes to this richer setting. Here, we will have to distinguish time-bounded analysis problems from others. With regard to the former, our idea is to employ the digitisation approach from [59] to derive upper and lower bounds for time-bounded reachability probabilities. In the unbounded case, it will be possible to completely abstract from the continuous behaviour of the Markov Automaton by instead considering the underlying Markov Decision Process.

6.5 Conclusion

To tackle the problem of correctness and reliability of control software in the aerospace domain, *formal methods* are increasingly being employed. They enable the exhaustive and mathematically founded analysis of all possible behaviours of a computer program and of its interaction with the overall system and the verification of properties such as functional correctness. They also allow to reduce the effort and, thus, the cost of testing activities [4]. Due to their benefits, they are increasingly becoming an integral part of the development cycle of safety-critical systems [44, 47, 62].

We have given a sketch of the ESA-funded COMPASS project, the related toolset and its underlying techniques. COMPASS provides an integrated approach to integrated system-software co-engineering covering modelling, analysis, and verification activities. While these methods turned out to be very useful in practical applications, there is still room for technological improvements. In the second part of this chapter, we have identified current bottlenecks and possible solutions. This comprises techniques that cover both the nominal and the error behaviour of systems, such as the formal validation of quantitative requirement specifications (Sect. 6.4.1) and multi-objective verification (Sect. 6.4.7).

Other methods focus on fault management, with the goal of improving the expressivity of error modelling and related analysis methods. In this category, we find approaches such as contract-based failure analysis (Sect. 6.4.2), Timed Failure Propagation Graphs (Sect. 6.4.4), parametric error models (Sect. 6.4.5), and Dynamic Fault Trees (Sect. 6.4.6). Last but not least, there is demand for better support for fault diagnosis and management in the form of FDIR design (Sect. 6.4.3).

To guarantee a smooth embedding of such technologies in the overall system development process, additional support by accompanying process-oriented measures is required. Here, the main concern is the integration of the modelling, analysis, and validation activities enabled by COMPASS with the design and implementation steps as provided by other tools supporting AADL (such as TASTE [78]) or other specification languages (such as Simulink). Moreover we note that our approach is completely model based. Thus, methods for generating code from AADL specifications and for checking the conformance of a hardware/software implementation with respect to the AADL model are required. For the latter, model-based testing [30] can be employed, which is an automated technique in which the test generation process is steered by the AADL model.

References

1. S. Abdelwahed, G. Karsai, N. Mahadevan, S. Ofsthun, Practical implementation of diagnosis systems using timed failure propagation graph models. IEEE Trans. Instrum. Meas. **58**(2), 240–247 (2009)
2. E. Alaña, H. Naranjo, Y. Yushtein, M. Bozzano, A. Cimatti, M. Gario, R. de Ferluc, G. Garcia, Automated generation of FDIR for the COMPASS integrated toolset (AUTOGEF), in *Proceedings of DASIA 2012*, vol. ESA SP 701 (2012)
3. J. Alonso, M. Grottke, A.P. Nikora, K.S. Trivedi, An empirical investigation of fault repairs and mitigations in space mission system software, in *Proceedings of DSN 2013* (IEEE, 2013), pp. 1–8
4. P. Anderson, Detecting bugs in safety-critical code. Dr. Dobb's J. **33**(3), 22–27 (2008), http://www.drdobbs.com/tools/detecting-bugs-in-safety-critical-code/206104422
5. M. Autili, L. Grunske, M. Lumpe, P. Pelliccione, A. Tang, Aligning qualitative, real-time, and probabilistic property specification patterns using a structured English grammar. IEEE Trans. Software Eng. **41**(7), 620–638 (2015)
6. C. Baier, B. Haverkort, H. Hermanns, J.P. Katoen, Model-checking algorithms for continuous-time Markov chains. IEEE Trans. Software Eng. **29**(6), 524–541 (2003)
7. C. Baier, B.R. Haverkort, H. Hermanns, J.P. Katoen, Model checking meets performance evaluation. SIGMETRICS Perform. Eval. Rev. **32**(4), 10–15 (2005)
8. C. Baier, J.P. Katoen, *Principles of Model Checking* (MIT Press, New York, 2008)
9. E. Bartocci, R. Grosu, P. Katsaros, C.R. Ramakrishnan, S.A. Smolka, Model repair for probabilistic systems, in *Proceedings of TACAS 2011*. LNCS, vol. 6605 (Springer, 2011), pp. 326–340
10. A. Biere, A. Cimatti, E. Clarke, Y. Zhu, Symbolic model checking without BDDs, in *Proceedings of TACAS 1999*. LNCS, vol. 1579 (Springer, 1999), pp. 193–207
11. A. Biere, K. Heljanko, T.A. Junttila, T. Latvala, V. Schuppan, Linear encodings of bounded LTL model checking. Logical Methods Comput. Sci. **2**(5) (2006)
12. B. Bittner, Formal failure analyses for effective fault management: an aerospace perspective, Ph.D. thesis, University of Trento, 2016
13. B. Bittner, M. Bozzano, R. Cavada, A. Cimatti, M. Gario, A. Griggio, C. Mattarei, A. Micheli, G. Zampedri, The xSAP safety analysis platform, in *Proceedings of TACAS 2016*. LNCS, vol. 9636 (Springer, 2016), pp. 533–539
14. B. Bittner, M. Bozzano, A. Cimatti, Automated synthesis of timed failure propagation graphs, in *Proceedings of IJCAI 2016* (AAAI Press, 2016), pp. 972–978
15. B. Bittner, M. Bozzano, A. Cimatti, R. de Ferluc, M. Gario, A. Guiotto, Y. Yushtein, An integrated process for FDIR design in aerospace, in *Proceedings of IMBSA 2014*. LNCS, vol. 8822 (Springer, 2014), pp. 82–95
16. B. Bittner, M. Bozzano, A. Cimatti, X. Olive, Symbolic synthesis of observability requirements for diagnosability, in *Proceedings of AAAI-12* (2012)
17. B. Bittner, M. Bozzano, A. Cimatti, G. Zampedri, Automated verification and tightening of failure propagation models, in *Proceedings of AAAI 2016* (2016), pp. 3724–3730
18. V. Bos, H. Bruintjes, S. Tonetta, Catalogue of system and software properties, in *Proceedings of SAFECOMP 2016*. LNCS, vol. 9922 (Springer, 2016), pp. 88–101
19. H. Boudali, P. Crouzen, M. Stoelinga, A rigorous, compositional, and extensible framework for dynamic fault tree analysis. IEEE Trans. Dependable Secure Comput. **7**(2), 128–143 (2010)

20. M. Bozzano, R. Bruttomesso, A. Cimatti, T. Junttila, P. van Rossum, S. Schulz, R. Sebastiani, Mathsat: tight integration of SAT and mathematical decision procedures. J. Autom. Reason. **35**, 265–293 (2005)
21. M. Bozzano, A. Cimatti, M. Gario, A. Micheli, SMT-based validation of timed failure propagation graphs, in *Proceedings of AAAI 2015* (2015), pp. 3724–3730
22. M. Bozzano, A. Cimatti, M. Gario, S. Tonetta, Formal design of fault detection and identification components using temporal epistemic logic, in *Proceedings of TACAS 2014*. LNCS, vol. 8413 (Springer, 2014), pp. 46–61
23. M. Bozzano, A. Cimatti, M. Gario, S. Tonetta, Formal design of asynchronous fault detection and identification components using temporal epistemic logic. Logical Methods Comput. Sci. **11**(4), 1–33 (2015)
24. M. Bozzano, A. Cimatti, J.P. Katoen, P. Katsaros, K. Mokos, V.Y. Nguyen, T. Noll, B. Postma, M. Roveri, Spacecraft early design validation using formal methods. Reliab. Eng. Syst. Safety **132**, 20–35 (2014)
25. M. Bozzano, A. Cimatti, J.P. Katoen, V.Y. Nguyen, T. Noll, M. Roveri, Safety, dependability, and performance analysis of extended AADL models. Comput. J. **54**(5), 754–775 (2011)
26. M. Bozzano, A. Cimatti, C. Mattarei, A. Griggio, Efficient anytime techniques for model-based safety analysis, in *Proceedings of CAV 2015*. LNCS, vol. 9206 (Springer, 2015), pp. 603–621
27. M. Bozzano, A. Cimatti, C. Mattarei, S. Tonetta, Formal safety assessment via contract-based design, in *Proceedings of ATVA 2014* (2014), pp. 81–97
28. M. Bozzano, A. Cimatti, F. Tapparo, Symbolic fault tree analysis for reactive systems, in *Proceedings of ATVA 2007*. LNCS, vol. 4762 (Springer, 2007), pp. 162–176
29. T. Brázdil, V. Forejt, J. Kretínský, A. Kucera, The satisfiability problem for Probabilistic CTL, in *Proceedings of LICS 2008* (IEEE, 2008), pp. 391–402
30. M. Broy, B. Jonsson, J.P. Katoen, M. Leucker, A. Pretschner, (eds.), *Model-Based Testing of Reactive Systems: Advanced Lectures*. LNCS, Vol. 3472 (Springer, 2005)
31. H. Bruintjes, J.P. Katoen, D. Lemens, A statistical approach for timed reachability in AADL models, in *Proceedings of DSN 2015* (IEEE CS Press, 2015), pp. 81–88
32. R. Cavada, A. Cimatti, M. Dorigatti, A. Griggio, A. Mariotti, A. Micheli, S. Mover, M. Roveri, S. Tonetta, The nuXmv symbolic model checker. CAV **2014**, 334–342 (2014)
33. S. Chakraborty, J.P. Katoen, On the satisfiability of some simple probabilistic logics, in *Proceedings of LICS 2016* (ACM, 2016), pp. 56–66
34. A. Cimatti, E. Clarke, E. Giunchiglia, F. Giunchiglia, M. Pistore, M. Roveri, R. Sebastiani, A. Tacchella, NuSMV 2: an open-source tool for symbolic model checking, in *Proceedings of CAV 2002*. LNCS, vol. 2404 (Springer, 2002), pp. 359–364
35. A. Cimatti, R. Demasi, S. Tonetta, Tightening a contract refinement, in *Proceedings of SEFM 2016* (2016), pp. 386–402
36. A. Cimatti, M. Dorigatti, S. Tonetta, OCRA: a tool for checking the refinement of temporal contracts, in *Proceedings of ASE 2013* (2013), pp. 702–705
37. A. Cimatti, C. Pecheur, R. Cavada, Formal verification of diagnosability via symbolic model checking, in *Proceedings of IJCAI 2003* (Morgan Kaufmann, 2003), pp. 363–369
38. A. Cimatti, S. Tonetta, Contracts-refinement proof system for component-based embedded systems. Sci. Comput. Program. **97**, 333–348 (2015)
39. The COMPASS project, http://www.compass-toolset.org/
40. COMPASS user manual. Technical Report. Version 3.0, COMPASS Consortium (2016), http://www.compass-toolset.org/docs/compass-manual.pdf
41. COMPASS tutorial. Technical Report Version 3.0, COMPASS Consortium (2016), http://www.compass-toolset.org/docs/compass-tutorial.pdf
42. C. Dehnert, S. Junges, N. Jansen, F. Corzilius, M. Volk, H. Bruintjes, J.P. Katoen, E. Abraham, PROPhESY: a probabilistic parameter synthesis tool, in *Proceedings of CAV 2015*, LNCS, vol. 9206 (Springer, 2015), pp. 214–231
43. S. Derisavi, H. Hermanns, W.H. Sanders, Optimal state-space lumping in Markov chains. Inf. Process. Lett. **87**(6), 309–315 (2003)

44. Software considerations in airborne systems and equipment certification. Software Standard DO-178C/ED-12C, RTCA Inc. and EUROCAE (2011)
45. J.B. Dugan, S.J. Bavuso, M.A. Boyd, Dynamic fault-tree models for fault-tolerant computer systems. IEEE Trans. Reliab. **41**(3), 363–377 (1992)
46. M. Dwyer, G. Avrunin, J. Corbett, Patterns in property specifications for finite-state verification, in *Proceedings of ICSE 1999* (IEEE CS Press, 1999), pp. 411–420
47. Space engineering: Verification. ECSS Standard E-ST-10-02C, European Cooperation for Space Standardization (2009)
48. Space engineering: System engineering general requirements. ECSS Standard E-ST-10C, European Cooperation for Space Standardization (2009)
49. Space product assurance: Failure modes, effects (and criticality) analysis (FMEA/FMECA). ECSS Standard Q-ST-30-02C, European Cooperation for Space Standardization (2009)
50. Space product assurance: Availability analysis. ECSS Standard Q-ST-30-09C, European Cooperation for Space Standardization (2008)
51. Space product assurance: Dependability. ECSS Standard Q-ST-30C, European Cooperation for Space Standardization (2009)
52. Space product assurance: Fault tree analysis—adoption notice ECSS/IEC 61025. ECSS Standard Q-ST-40-12C, European Cooperation for Space Standardization (2008)
53. Space product assurance: Safety. ECSS Standard Q-ST-40C, European Cooperation for Space Standardization (2009)
54. M.A. Esteve, J.P. Katoen, V.Y. Nguyen, B. Postma, Y. Yushtein, Formal correctness, safety, dependability and performance analysis of a satellite, in *Proceedings of ICSE 2012* (ACM and IEEE CS Press, 2012), pp. 1022–1031
55. K. Etessami, M.Z. Kwiatkowska, M.Y. Vardi, M. Yannakakis, Multi-objective model checking of Markov decision processes. Logical Methods Comput. Sci. **4**(4) (2008)
56. V. Forejt, M. Kwiatkowska, D. Parker, Pareto curves for probabilistic model checking, in *Proceedings of ATVA 2012*. LNCS, vol. 7561 (Springer, 2012), pp. 317–332
57. M. Gario, A formal foundation of FDI design via temporal epistemic logic. Ph.D. thesis, Trento University, Italy (2016), https://marco.gario.org/phd/gario_phd.pdf
58. D. Guck, T. Han, J.P. Katoen, M.R. Neuhäußer, Quantitative timed analysis of interactive Markov chains, in *Proceedings of NFM 2012*. LNCS, vol. 7226 (Springer, 2012), pp. 8–23
59. D. Guck, H. Hatefi, H. Hermanns, J.P. Katoen, M. Timmer, Modelling, reduction and analysis of Markov automata, in *Proceedings of QEST 2013*. LNCS, vol. 8054 (Springer, 2013), pp. 55–71
60. K. Heljanko, T.A. Junttila, T. Latvala, Incremental and complete bounded model checking for full PLTL, in *Proceedings of CAV 2005*. LNCS, vol. 3576 (2005), pp. 98–111
61. H. Hermanns, *Interactive Markov Chains: The Quest for Quantified Quality*. LNCS, vol. 2428 (Springer, 2002)
62. G.J. Holzmann, The power of 10: rules for developing safety-critical code. Computer **39**(6), 95–99 (2006)
63. N. Jansen, F. Corzilius, M. Volk, R. Wimmer, E. Abraham, J.P. Katoen, B. Becker, Accelerating parametric probabilistic verification, in *Proceedings of QEST 2014*. LNCS, vol. 8657 (Springer, 2014), pp. 404–420
64. S. Junges, D. Guck, J.P. Katoen, A. Rensink, M. Stoelinga, Fault trees on a diet, in *Proceedings of SETTA 2015*. LNCS, vol. 9409 (Springer, 2015), pp. 3–18
65. J.P. Katoen, V.Y. Nguyen, T. Noll, Formal validation methods in model-based spacecraft systems engineering, in *Modeling and Simulation-Based Systems Engineering Handbook*, Chap. 14 (CRC Press, 2014), pp. 339–375
66. J.P. Katoen, L. Song, L. Zhang, Probably safe or live, in *Proceedings of CSL-LICS 2014* (ACM, 2014), pp. 55:1–55:10
67. J.P. Katoen, I.S. Zapreev, E.M. Hahn, H. Hermanns, D.N. Jansen, The ins and outs of the probabilistic model checker MRMC. Perform. Eval. **68**(2), 90–104 (2011)
68. M. Kwiatkowska, G. Norman, D. Parker, H. Qu, Compositional probabilistic verification through multi-objective model checking. Inf. Comput. **232**, 38–65 (2013)

69. MathSAT, http://mathsat.fbk.eu
70. A. Misra, J. Sztipanovits, A. Underbrink, R. Carnes, B. Purves, Diagnosability of dynamical systems, in *3rd International Workshop on Principles of Diagnosis* (1992), pp. 239–244
71. MRMC – Markov Reward Model Checker, http://www.mrmc-tool.org/
72. T. Noll, Safety, dependability and performance analysis of aerospace systems, in *Proceedings of FTSCS 2014*. CCIS, vol. 476 (Springer, 2015), pp. 17–31
73. Nonelectronic parts reliability data (NPRD-2016). Technical Report, Quanterion Solutions Inc. (2015), https://www.quanterion.com/product/publications/nonelectronic-parts-reliability-data-publication-nprd-2016/
74. The NuSMV model checker, http://nusmv.fbk.eu
75. The nuXmv model checker, https://nuxmv.fbk.eu/
76. S.C. Ofsthun, S. Abdelwahed, Practical applications of timed failure propagation graphs for vehicle diagnosis, in *Proceedings of Autotestcon 2007* (IEEE, 2007), pp. 250–259
77. S. Pathak, E. Abraham, N. Jansen, A. Tacchella, J.P. Katoen, A greedy approach for the efficient repair of stochastic models, in *Proceedings of NFM 2015*. LNCS, vol. 9058 (Springer, 2015), pp. 295–309
78. M. Perrotin, E. Conquet, J. Delange, A. Schiele, T. Tsiodras, TASTE: a real-time software engineering tool-chain overview, status, and future, in *Proceedings of SDL 2011*. LNCS, vol. 7083 (Springer, 2012), pp. 26–37
79. I. Pill, S. Semprini, R. Cavada, M. Roveri, R. Bloem, A. Cimatti, Formal analysis of hardware requirements, in *Proceedings of DAC 2006* (2006), pp. 821–826
80. Reliability Prediction of Electronic Equipment. No. MIL-HDBK-217F in Military standardization handbook. Department of Defense, USA (1995), http://quicksearch.dla.mil/qsDocDetails.aspx?ident_number=53939
81. Architecture Analysis & Design Language (AADL) Annex, Volume 1, Annex E: Error Model Annex. SAE Standard AS5506/1A (International Society of Automotive Engineers, 2015)
82. Architecture Analysis and Design Language (AADL) Annex, Volume 1, Annex A: Graphical AADL Notation. SAE Standard AS5506/1 (International Society of Automotive Engineers, 2006)
83. Architecture Analysis & Design Language (AADL). SAE Standard AS5506 (International Society of Automotive Engineers, 2004)
84. Architecture Analysis & Design Language (AADL) (rev. B). SAE Standard AS5506B (International Society of Automotive Engineers, 2012)
85. Solar Orbiter, http://sci.esa.int/solar-orbiter/
86. A. Valmari, G. Franceschinis, Simple $O(m \log n)$ time Markov chain lumping, in *Proceedings of TACAS 2010*. LNCS, vol. 6015 (Springer, 2010), pp. 38–52
87. M. Volk, S. Junges, J.P. Katoen, Advancing dynamic fault tree analysis – get succinct state spaces fast and synthesise failure rates, in *Proceedings of SAFECOMP 2016*. LNCS, vol. 9922 (Springer, 2016), pp. 253–265
88. R. Wimmer, M. Herbstritt, H. Hermanns, K. Strampp, B. Becker, Sigref – a symbolic bisimulation tool box, in *Proceedings of ATVA 2006*. LNCS, vol. 4218 (Springer, 2006), pp. 477–492

Printed in the United States
By Bookmasters